The Authors

Sh. L.K. Jain acquired his master degree in Agronomy from Rajasthan College of Agriculture (SKRAU, Bikaner), Udaipur with Gold Medal in 1999. Presently, he is associated in teaching UG students since last five years. He has contributed in publication of 3 books, 13 chapters and a number of research papers in national and international repute journals. He has a vast experience of research as SRF in AICRPDA and extension services as SMS in KVK Barmer.

Dr. Ravindra Kumar Jain is presently working as Senior Research Fellow in organic farming project at Directorate of Research, Maharana Pratap University of Agriculture and Technology, Udaipur. He obtained Ph.D. degree from MPUAT Udaipur, specializing in Molecular Biology. He has to his credit 17 research papers in National and International Journals and attended 20 National and International Seminar / Symposia.

Dr. Hanuman Prasad Parewa a BHU alumnus is presently the faculty member at College of Agriculture (Agriculture University, Jodhpur) Sumerpur, Pali, Rajasthan. He got UGC scholarship for his Ph. D. programme in Soil Science & Agricultural Chemistry. He has five year of experience in undergraduate teaching in the field of soil science and more than 2 years of experience in extension and administration. He has published 10 papers, 5 book chapters, 17 popular articles, 5 folder and about 15 abstract and full papers including the proceeding of National and International seminars and conferences. He has specialized in Soil fertility, INM and PGPR.

Dr. Sawai Dan Ratnoo, obtained his Ph.D. degree from the SKRAU Bikaner in 1995 in the field of Entomology. He has 25 years research, extension and teaching experience in the field of agriculture particularly in insect pest management of crops and stored grain pests. He has published many packages of practices for Zone II b of Rajasthan as an authority of Zonal Director Research. He also published more than 75 research paper and has actively participated in many national and international scientific conferences and seminar. He is the member of several organizations and societies. He is presently working as Dean and Faculty Chairman, Faculty of Agriculture, Agriculture University Jodhpur besides Zonal Director Research ARS, Keshwana, Jalore under jurisdiction of AU Jodhpur.

Manual on Fundamentals of Agronomy

Lokesh Kumar Jain

Assistant Professor (Agronomy)
College of Agriculture
Sumerpur

Ravindra Kumar Jain

Senior Research Fellow
Directorate of Research, MPUAT Udaipur

Hanuman Prasad Parewa

Assistant Professor (Soil Science and Ag. Chemistry)
College of Agriculture
Sumerpur

Sawai Dan Ratnoo

Dean & Faculty Chairman
College of Agriculture
Sumerpur

SCIENTIFIC PUBLISHERS

Published by
SCIENTIFIC PUBLISHERS (INDIA)

5 A, New Pali Road
P.O. Box 91
Jodhpur - 342 001 INDIA

ISBN: 978-93-88172-01-1

Visit the Scientific Publishers (India) website at
http://www.scientificpub.com
Printed in India

Prof. Balraj Singh
Vice Chancellor

Agriculture University
Jodhpur - 342 001

FOREWORD

The subject Agronomy deals with the principles and practices of crop production and soil management and an essential need is being realized that the students of Agriculture must have a sound knowledge of practical aspects of Agronomy. The manual on Fundamentals of Agronomy places a premium on ideas or perspectives both of an academic and students as it has wide coverage and cover all the prescribed topics such as identifications of crops, seeds, weeds, fertilizers and plant protection chemicals, weed management, water quality analysis and measurement. The authors approach to compile the experiments in a new form as suggested by Fifth Dean Committee, to enrich and strengthen the study wherever possible is highly appreciable. It is felt that this book meets a long felt need of students.

I congratulate Sh. L.K. Jain, Dr. R.K. Jain, Dr. H.P. Parewa and Dr. S.D. Ratnoo for scrupulous efforts in bringing out a Manual on Fundamentals of Agronomy. As it covers a core course totally modified recently and implemented during current session of 2017-18 for undergraduates in all Agriculture Universities in the country and it will serve a very useful purpose in this direction.

(Prof. Balraj Singh)

Preface

This book is intended as a text for undergraduate students of Agriculture. It also be useful to research scholars and other professionals in the field of agriculture development and management especially under teaching stream.

Introductory Agronomy involves several basic subjects like agronomy, soil and water, farm machinery, engineering, soil science and plant breeding and genetics etc. For an integrated development and management of agriculture knowledge of all these subjects are necessary for undergraduate students. A sincere attempt is made to provide such prospective to the students.

A fundamental knowledge of identification of crops, seeds, weeds, fertilizers and plant protection chemicals, water quality analysis and measurement will be needed in crop planning under different situations. Therefore, an attempt has been to present the topics relevant to the needs of the agronomy. Thus, book is therefore, designed to fulfill the need for students of agriculture and serves as reference tool for the teachers in the field of Agronomy from all points of view.

A collage of material and text from different sources has been used to prepare this text book. The authors acknowledge their indebtedness to authors of books, manuals and internet media from which most of the materials have been drawn. In most of cases, it has obtained permission of reproduction, for which the authors and the publisher offer their apologies. At the end of the book and at footnotes of particular figures, references have been given from which most of the material has been taken.

The authors express their sincere gratitude to Hon'ble Vice Chancellor Prof, Balraj Singh and thanks to Dr. I.S. Naruka, Dr. R.L. Bhardwaj and other colleagues for their valuable suggestions in improving the quality from time to time during preparation of first edition of Fundamentals of Agronomy. The authors are highly acknowledged to our all well wishers for their direct-indirect cooperation.

Last, but not the least, the authors would welcome critical suggestions if any received from the learned readers would be highly appreciated for further modifications and improvements in the text.

Authors

Contents

1

IDENTIFICATION OF CROPS AND THEIR SEEDS

Introduction

The identification of different crops and their seeds with knowledge of botany and visible characters is important for agriculture students. In systematic botany or taxonomy the closely related or similar type of plants are grouped into a single category and are: family, genus, species etc. In identification, the particular crop or seed in question must be identified up to the species level. There are so many field crops grown throughout the country in *Kharif, Rabi, Spring* and *Summer* season in respect of consumption, utilization, economics, productivity etc. In general, crop is defined as a group of identical plants grown in field for economic importance or aggregation of individual plant species grown in an unit area for economic purpose. Plants can usually be grouped together in respect of agronomic/commercial/economic classification such as cereals, pulses, oilseeds, fibre, sugar, tuber and fodder crops and identified on the basis of morphological description like root, stem, leaves, inflorescence and other basic characteristics. The seed is defined as a fertilized mature ovule and consists of an embryo, a protective covering (seed coat) and stored food (endosperm). The grain which is used for multiplication is called as seed while those used for human/animal consumption are called as grains. The identification of seed is usually by comparison, comparing the seeds with a mental image of what something should be, with specimens in a reference collection or with illustration of seeds. In most cases, the useful clues for the identification of seed came from the following characters:

- The size, shape and color of seeds
- The nature, arrangement and pattern of markings like lines, ridges, pits, projection on the seed surface
- The shape and position of the attachment scar
- The presence of wings, hair or scale, spines etc

- The internal structure, position and size of the embryo, presence or absence of the endosperm

Once the seed is characterized for a particular family, identification could easily be done by studying the above mentioned seed characters. Seeds particularly of an unknown and unconventional crop and weed are difficult to identify as such, growing it as a plant could help in identification of seeds.

The pictures or physical appearance of different crops have been depicted in Sheet 1 to 3.

Identification of crop plants

Some of the important crops have been summarized as under:

Wheat

It is a major *rabi/* winter crop and second staple food of Indians. The plant normally attains a height of 80-120 cm.

Root system: The plants have primary root system i.e. the roots arise when seed germinates at the depth where it is placed and secondary root system i.e. it arises at a point above the primary root system and act as the principal organ of absorption as the young seedling progresses to maturity.

Shoot system: It comprises of all plant parts visible above the ground and composed of stem, leaves and inflorescence.

i. Stem: The stem of wheat plant is round or cylindrical and solid at nodes. Stem may be called as culm.

ii. Leaves: Leaf consists of four parts.

a. **Leaf sheath:** It is the basal part of leaf, encircles the (culm) stem and protects the growing point and auxillary buds from the adverse weather and provides support to stem to some extent.

b. **Leaf blade:** The flattened, parallel veined portion of the leaf.

c. **Ligule***:* A membranous or cartilaginous fringe at the junction of the sheath and blade on the side of leaf next to culm. The continuation of the sheath through the collar is known as ligule.

d. **Auricle:** Lobes of leaf blade which extend downward on each side at the junction of the blade and sheath. These are horn or claw like appendages projecting from collar of leaf.

iii. Inflorescence: The flowering portion of wheat plant which is called ear or head or spike.

a. **Rachis***:* The central zigzag axis is the rachis. Spikelets are borne on alternate sides of rachis, which gives it zigzag appearance.

b. **Spikelet***:* It is composed of flowers called florets. The number of florets in a spikelet may vary from 1 to 6.

c. **Florets:** The outer covering of a floret is made up of a lemma and palea. The lemma enfolds the palea near their attachment point. If an awn is present, it is attached to lemma.

iv. **Kernel:** Wheat plant has a caryopsis type of fruit. The typical wheat kernel is from 3 to 10 mm in length and from 3 to 5 mm in diameter.

Maize

Root system: The root system of maize is deep and fibrous. It consists of:

i. **Seminal or temporary roots***:* They consist of radicle and number of lateral roots which arise at the base of the first node of the stem under soil surface just above the scutellar node.

ii. **Crown or coronal roots***:* They arise from the basal portion of the stem.

iii. **Brace, prop or aerial roots:** They arise from second, third and sometimes fourth node above the soil surface. All may or may not enter the soil.

Shoot system

i. **Stem:** The stem is made up of approximately 12-18 alternating nodes and internodes, and is completely filled with the pith. The number of internodes may vary but on an average there are 14 internodes. A leaf is attached to each node, and often a bud or branch arises at a node. Internodes are somewhat flattened or grooved on the side next to the leaf sheath.

ii. **Leaves:** The leaves of maize develop alternately on opposite sides of the stem. Each leaf consists of a thin, flat and expanded blade with definite midrib and smaller veins and a thicker, more rigid sheath. Each sheath surrounds the internode above the node to which it is attached. The number of leaves varies from 12 to 20 and stomata are present on both surfaces of the leaf.

iii. **Inflorescence:** Maize is a monoecious plant having both male and female flowers on the same plant. Male flower is borne in a tassel at the top of stem and female flower is borne inside the young cob.

iv. **Kernel or caryopsis:** Maize kernel is one seeded fruit or caryopsis. Seed enclosed within the pericarp consists of the embryo, endosperm and remnants of seed coat and nucleus.

Sorghum

Root system: Roots are finer and more fibrous than maize. Most of the roots are confined to the upper 15 cm of soil but they may reach a depth of one or one and half meter. Prop roots may develop from the auxiliary buds on the lowest node of stem above the ground level.

Shoot system

i. **Stem:** Sorghum stems are solid, though the centre may become spongy with spaces in the pith. Culms have 7-18 nodes and internodes.

ii. **Leaves:** The number of leaves on the main stem varies from 7 to 24 having a length of 30 to 135 cm, and a width of between 1.5 and 13 cm at the widest point. The length of the leaf sheath may vary between 15 and 35cm.

iii. **Inflorescence:** The inflorescence of sorghum is a panicle and also known as head. The panicle is composed of spikelets which occur in pairs, one of them being sessile and the other pedicellate. Sessile spikelet has two glumes. The glumes enclose two florets, the upper being perfect, the lower sterile and consisting of lemma only. There are two lodicules lying adjacent to base of the spikelet.

Pearl millet

Root system: It is fibrous consisting of primary, secondary and prop roots.

Shoot system

i. **Stem:** Solid and single, made up of nodes and internodes. Nodes are swollen while internodes are cylindrical and glabrous.

ii. **Leaves:** Leaves are long and faity broad and held erect by thick midribs. Leaves consist of blades and sheath. The length of spike varies from 15 to 60 cm. Spikelets occur in pairs.

iii. **Inflorescence:** Pearlmillet inflorescence is a compound terminal spike called panicle and its length generally varies between 20-25 cm with a circumference of 7-9 cm. The panicle shape varies considerably, but, the common shapes are cylindrical or conical.

iv. **Caryopsis (grain):** The seed is 3-4 mm long and 2-2.5 mm broad. The colour of grain varies from whitish yellow to grey or dull light blue.

Barley

Root system: It consists of shallow and deep roots. Shallow roots spread laterally. Deep roots go up to 7.5 to 150 cm.

Shoot system

i. **Stem:** Stem is cylindrical and possesses five to seven hollow internodes separated by solid nodes.

ii. **Leaves:** Leaves arise alternately and consist of a sheath, blade, ligule and auricle. Auricles are very conspicuous.

iii. **Inflorescence:** Inflorescence is called spike or head. Spike consists of spikelets attached at the nodes of a zigzag rachis, and each spikelet has two glumes and a floret. In two rowed barley, only central spikelet is fertile whereas in six rowed barley all the three spikelets are fertile.

iv. **Kernel (grain):** Its grain is a caryopsis consisting of lemma, palea and a rachilla. Caryopsis is composed of the pericarp, endosperm and embryo.

Chickpea

Root system: It is robust, up to 2 m deep, but major portion of the roots is confined up to 60 cm depth.

Shoot system

i. **Stem:** The plant possesses branched, erect or spreading, sometimes shrubby much branched stems, which are 0.2-1.0 m tall, glandular pubescent, olive, dark green or bluish green in color

ii. **Leaves:** They are imparipinnate, glandular-pubescent with 3-8 pairs of leaflets and a top leaflet (rachis ending in a leaflet); leaflets ovate to elliptical, 0.6-2.0 cm long, 0.3-1.4 cm wide; margin serrated, apex acuminate to aristate, base cuneate; stipules 2-5 toothed and stipules are absent.

iii. **Flowers:** Flowers are solitary, sometimes two per inflorescence, auxiliary; peduncles 0.6-3.0 cm long, pedicels 0.5-1.3 cm long, bracts triangular or tripartite; calyx 7-10 mm long; corolla white, pink, purplish (fading to blue) or blue, 0.8-1.2 cm long. The staminal column is diadelphous (9-1) and the ovary is sessile, inflated and pubescent (generally, one to two seeds and having a maximum of three seeds).

iv. **Pods:** Pod is rhomboid ellipsoid (having 1-2 with three seeds as a maximum), inflated and glandular-pubescent. Seed color cream, yellow, brown, black or green, rounded to angular, seed coat smooth or wrinkled, or tuberculate, laterally compressed with a median groove around two-third of the seed, anterior beaked.

Lentil

Herbaceous annual plant, erect and bushy with four to six primary branches.

Root system: It has well developed root system including a central tap root and many lateral branches.

Shoot system

i. **Stem:** Stem is weak and quadrangular.

ii. **Leaves:** Leaves are small, compound and pinnate, small sessile leaflets occur in pairs of five to seven. The end of leaflets forms tendrils.

iii. **Inflorescence:** It is a raceme of two to four flowers. Flowers are small, white tinged with blue, violet or pink colour.

iv. **Pods:** Pods are short, flattened, 1.0 to 1.5 cm long with a curved beak.

Groundnuts

It is an annual plant having about 30-60 cm height.

Root system: It has a deep tap root system with well developed lateral roots.

Shoot system

i. **Stem:** Angular and hairy stem with spreading or erect branches.

ii. **Leaves (cataphylls):** Leaves are pinnate with two pairs of ovate leaflets.

iii. **Inflorescence:** Solitary or raceme and flowers are borne at the axils of the leaves, either above or below the ground.

iv. **Fruit:** The fruit is an indehiscent pod normally contains one to three seeds. The shell of pod which contains seed is morphologically the pericarp, and the thin skin that covers the seed is testa.

Sesame

It is an herbaceous annual growing to a height of 0.5 to 2.0 m.

Root system: In early maturing varieties, it is poorly developed while in late maturing varieties, it is well developed.

Shoot system

i. **Stem:** Stem is erect, normally square in section, darkish green coloured and covered with short hairs (upper part).

ii. **Leaves:** Leaves are variable in shape and size. Lower leaves are broader while the upper ones are narrow. The arrangement of leaves on the stem may be opposite, alternate or mixed.

iii. **Inflorescence:** The inflorescence is raceme and the flowers arise in the axils of the leaves and on the upper portion of the stem and branches.

iv. **Fruit:** Fruit is capsule having two or four carpels. Numerous seeds are borne in capsule. Colour of seed may be white, black and white brown.

Cotton

In the wild state, cotton is a permanent plant but most of the cultivated cottons are annual.

Root system Plant has a tap root with secondary roots that branch laterally from primary root.

Shoot system

i. **Stem:** The main stem is erect and much branched. The branches develop from buds located at the nodes of main stem. There are two types of branches:

(a) **Monopodial branches:** The true auxiliary bud develops into a vegetative branch which only bears leaves and no flowers. This is known as monopodial branch.

(b) **Sympodial/fruiting branch:** The auxillary bud generally develops into sympodial or fruiting branch which bears flowers. Lower branches are usually vegetative and upper ones are fruiting.

ii. **Leaves:** Leaves are spirally arranged on the main stem and vegetative branches. The leaves buds which appear as small, pyramid shaped, green structure are called as square. Square consists of three triangular-shaped leafy structures known as "bracteoles" and the flower bud.

iii. **Fruit:** Fruit is the enlarged ovary that develops into three to five loculed capsule or boll. Bolls vary in shape and size but are more or less egg shaped. Number of seeds in each boll may be 24-50.

Sugarcane

It is a tall perennial plant growing erect up to 5 to 6 m height.

Root system: It is fibrous and consists of sett roots and shoot roots.

i. **Sett roots:** When sugarcane sett is planted in moist soil, the root primordia situated at the base of every cane joint is activated and produces temporary roots called as sett roots.

ii. **Shoot roots:** After the emergence of primary shoot from the bud, other roots are produced from lower rings of the lower nodes of the shoot. Those formed first go downwards, whereas those formed near the soil surface grow in upper layer of soil for providing anchorage for the plant. These roots are known as shoot roots.

Shoot system

i. **Stalk:** Sugarcane stalk is roughly cylindrical and is composed of many distinct nodes and internodes. It is above ground portion of plant which bears leaves and flowers.

ii. **Eye/bud:** At each node, there is a bud, sometime known as "eye" appearing on opposite sides of cane. These buds are protected by leaf sheath.

iii. **Leaf scar:** Just below the bud is a raised portion known as the leaf scar – a point of attachment of leaf sheath to the stalk.

iv. **Tillers:** Sugarcane produces branches below the soil surface. The underground portion of stem tapers rapidly and from the lateral buds of this region the shoots develop. Single cane may produce as many as 20-40 tillers.

v. **Leaves:** The leaves of cane plant grow alternately on opposite sides of the cane stalk from the nodes. Leaf consists of a sheath and the blade with a ligule in between.

vi. **Inflorescence:** The inflorescence of sugarcane normally called as arrow, is an open panicle. The arrangement of spikelets is racemose, that is, the oldest flowers are at the bottom and youngest at the top.

Seed characteristics of some common families

1. Poaceae: Cereals produce dry, one seeded fruits, called the "kernel" or "grain", in the form of a caryopsis, in which the fruit coat (pericarp) is strongly bound to the seed coat (testa). The embryo lies on outside of the endosperm and visible near the base of caryopsis on dorsal side.

2. Leguminoseae: Seeds vary largely in size, shape and surface characters. The fruit may be one seeded to several-seeded pod. Followings are important characteristics of sub family:

a. Mimosoideae' and Caesalpinodeae

i. The seeds are elongated broad and flattened, the two faces being plane or only rounded, colour is variable from black to white and yellow

ii. The hilum is very small, unspecialized and located at one end of the seed.

b. Papilinoideae

i. The seeds vary greatly in size, shape, colour and location of hilum and chalaza.

ii. In hilum, there is a fine longitudinal groove or slit down the middle. The area may be minute, as in some of the clovers, or may be large enough to be seen without magnification as in vetch.

iii. In some species the hilum is obscured by a persisting layer of corky tissue, as in cowpea and beans.

3. Crucifereae: The seeds unite, may be a true seed, in indehiscent pod or a segment of a pod.

i. The seeds are mostly spherical, or sometimes slightly flattened with reticulum or netting or lines or ridges on surface.

iii. The seed surface is covered with microscopic pits. These pits are usually covered with a whitish film, giving the appearance of white spots on the surface.

4. Polygonaceae: The fruit or so called seed is an achene which is three angled or flattened. The outer hull (pericarp) is hard, brown and glossy.

5. Chenopodiaceae: The seeds are flattened, circular or obvate in shape. The embryo is either in the form of a ring or horse shoe.

6. Caryophyllaceae: The seeds are black or brown, thick and flattened. The surface is roughened by tubercles of various types which are arranged in definite pattern.

7. Euphorbiaceae: Seeds vary largely in size, shape and surface configuration. The distinct feature of the seed is the presence of prominent raphe. The scar is a flattened area at the base. In some species the scar is obscured by caruncle (whitish corky outgrowth).

8. Solanaceae: The seeds are orbicular, oval or ovate with smooth, or variously configure width a reticulum, broken lines or pits on surface. They are more or less flattened and may be thick or thin. The embryo is curved with an abundant endosperm.

9. Compositeae: The seed unit is an indehiscent an achene which is one seeded fruit. The top of the achene is usually depressed. In many species there is a fringe of fine bristles or scales around the outer rim.

Identification characters of some crops' seeds

Rice: Rice grain is ripened ovary with lemma and palea firmly adheres to it. Rice fruit is a caryopsis in which single grain is fused with wall of the ovary (pericarp).

Wheat: The kernel has an amber color, and is larger and more tapered than hard red spring wheat. Kernels are long and pointed, usually lopsided and boat-shaped. The germ is protruding, oval and more pointed than hard red spring wheat. The crease is tight. A brush is not present on most varieties.

Barley: The grain is a caryopsis with lemma, palea and rachilla firmly adhere to it. Kernel shape is longer (8-12 mm long) and narrower, with more of a spike tooth taper at the end than in two-row barley. Two-thirds of the kernel (the outside two rows) is twisted, with a crooked crease. The crease is more open to the end.

Maize: Seed is large, flat and dented in the top. Color ranges from white, yellow, and red to strawberry. Seeds may have a white cap (Semi-dent, flint, sweet and pop corn will not be included in crop judging contests).

Sorghum: The seeds are more or less egg-shaped or oblong and somewhat flat. The color may be white, yellow, red or brown.

Pearlmillet: The grain is a caryopsis, almost oval in appearance with one of its ends tapering end that is attached to the spike. The colour of the grain varies from whitish yellow to dull light blue. The outer surface is smooth and shining.

Sesame: The fruit is a capsule which consists of numerous small ovate seeds of white, yellow, reddish brown or black colour.

Groundnut: The seeds of groundnut differ in size (length ranges from 7 to 21 mm with diameter from 5 to 13 mm), shape and colour of the seed coat or testa (red, pink, white etc.). The testa is thin and papery. In general, 3 unicellular layers viz., the outer epidermis or sclerenchyma, middle parenchyma and inner parenchyma constitute the testa.

Soybean: Seeds vary in color, and vary in shape from nearly round to oval. The color of the hilum (scar or spot on bean where it was attached to pod) varies from black to brown to tan or yellow and is a seed characteristic considered for identifying varieties.

Castor: The fruit is a spiny, greenish (to reddish-purple) capsule containing large, oval, shiny, bean-like seeds with variabie brownish mottling.

Rapeseed and mustard: The seeds are small round with smooth surface in yellow sarson (mustard) while rugose in brown *sarson* and *tori*. The seed colour may vary from yellow or brown depending upon variety.

Linseed (Flex): The seeds are flat and have a smooth, shiny surface. They usually are dark brown or yellow in color. The seeds are somewhat lens shaped, although more rounded at the base than the tip.

Safflower: Small hulled seed, light gray, tan or cream in color, often showing brown discoloration near point of attachment. Seeds are enclosed (15 to 30 seeds) in the flower head at maturity, which prevents shattering before harvest. Similar in shape to sunflower seed but of only half the size.

Sunflower: Seeds have a broad base but taper to a pointed end. The color is usually black or dark grey. An achene (shell) covers the nut-meat or seed within. They are similar to wild sunflower, only much larger and dark in color.

Gram: Seeds are characteristically 'beaked' or angular with a ridged or smooth seed coat. Seed colour varies from white to brown or black. Usually 1 to 3 seeds per pod.

Frenchbean: Seeds may be oblong, globular or kidney shaped varying in size, colur and shape according to variety. Colour may be white, yellow, pink, red purple, brown or black. Hilum is ringed with different colours. Pods usually bear upto 10 seeds per pod.

Greengram: The seeds are usually green but may be olive, yellow, brown or purplish brown in colour. Seeds are small, globular and flat hilum is white and nearly flat. Usually 7-11 seeds per pod.

Blackgram: The seeds are usually black but may be dark brown to green. Seeds are small, oblong, cylindrical to ovoid oblong, slightly truncated at ends measuring 4.0 to 5.2 mm long and 3.5 to 4.1 mm broad.

Moth: Seeds are small; light brown to dark reddish brown in colour with white hilum. Usually 4-9 seeds/pod are found.

Cowpea: The seeds are usually red, black, brown, green buff to white in colour. Seeds may be fully coloured or spotted, marbled, speckled, eyed or blotches. Seeds are 2.0-12.0 mm long, globular to renifrom smooth or wrinkled.

Horsegram: The seeds are ovoid, 5-8 seeds/pod, colour varies from light red, brown or black. Sometimes they may be faint mottles or with small scatterd black spot, hilum central.

Pigeonpea: Seeds are lenticular to ovoid, measuring upto 8mm in diameter and weight about 10 seeds/gram. Seeds are separated from each other by slight depression in pods.

Field pea: The seeds are round, angular or wrinkled in inflated or cylindrical pods usually 4-10 seeds per pod. They vary in color but are mostly yellow, grey or green.

Lentil: Seeds are "lens" shaped (round and rather flat). Seeds diameter ranges from 2.0 to 9.0 mm with light red to green red or brown testa with orange to yellow cotyledons. The seed surface is generally smooth, but on some large seeds may be wrinkled.

Cotton: The cotton fruits are called bolls which are usually more or less egg shaped and pointed at tips. These bolls are made up of several locules usually 3-5 and contain 7-11 seeds in each of them and each boll contains 24-50 seeds in it. Seeds are usually black or brown in colour with lint (long fiber) and fuzz (short fiber),

Sugarbeet: Seed is highly irregular in shape. The mature seed is contained within a mature reddish brown to brown seed coat. Seed are both multigerm and non-germ types.

Lucerne: Seeds vary in shape but often one half kidney shaped and one half mitt shaped and are greenish-yellow to light brown in color turning brown with age.

Berseem: Seeds are solitary and small in size. At maturity each floret contains one single seed. Seed is egg shaped, yellowish in colour and is around 2 mm in length.

Oat: The color is white, yellow or tan and the surface of the kernel is practically smooth. The awn, if present, is not bent or markedly twisted. The seed attachment is round and relatively small compared to the large sucker-mouth-shaped attachment in wild oats. Kernels are long and somewhat pointed at both ends, especially the tip end. The hull (the lemma and palea) is tightly attached to the kernel and accounts for 25 to 35 percent of its weight. The hulless kernel or groat comprises the remainder of the kernel weight.

Table 1: Agronomical information of different arable crops

Type of crop	Common Name	Scientific name	Family	Remarks
Cereal and millets	Wheat	*Triticum aestivum* L.	Poaceae	King of cereals
	Barley	*Hordeum vulgare* L.	Poaceae	Malting purpose
	Maize	*Zea mays* L.	Poaceae	Queen of cereals, Drosphila of crops
	Rice	*Oryza sativa* L.	Poaceae	Semi aquatic plant, Staple food of India
	Sorghum	*Sorghum bicolor* L.	Poaceae	Camel crop, King of coarse cereals
	Pearlmillet	*Pennisetum typhodium*	Poaceae	Poor man food
	Finger millet	*Eleusine coracena* Gaertn.	Poaceae	Minor millet, Famine reserve
	Proso millet	*Panicum miliacerum* L	Poaceae	Minor millet
	Barnyard millet	*Echinochloa frumentacea* L.	Poaceae	Minor millet
	Kodo	*Paspalum scrobiculatum* L.	Poaceae	Coarsest grain
	Kakun (Foxtail millet)	*Seteria italica* L.	Poaceae	Minor millet, Famine reserve
Pulses	Gram	*Cicer arietinum* L.	Fabaceae	King of pulses
	Lentil	*Lens esculenta* Moench	Fabaceae	Suitable for paira & utera cropping
	Pea	*Pisum sativum* L.	Fabaceae	Queen of pulses
	French bean	*Phaseolus vulgaris*	Fabaceae	Non-N fixing pulse
	Greengram	*Vigna radiate* L.	Fabaceae	High biological protein
	Blackgram	*Vign mungo* L.	Fabaceae	Highest phosphoric acid among pulses
	Moth	*Vigna acontifolia* L.	Fabaceae	Drought tolerant crop
	Cowpea	*Vigna sinensis* L.	Fabaceae	Vegetative meat, Good cover crop
	Arhar	*Cajanus cajan* L.	Fabaceae	Biological plough
	Horsegram	*Macrotiloma uniflorum* L.	Fabaceae	Arid legume
	Soybean	*Glycine max* L.	Fabaceae	Miracle crop, Wonder crop
	Clusterbean	*Cyamopsis tetragonoloba* L.	Fabaceae	Industrial crop for gum

Type of crop	Common Name	Scientific name	Family	Remarks
Oilseeds	Sarson	*Brassica compestris*	Cruciferae	Sarson
	Rai/indian mustard	*Brassica juncea*	Cruciferae	King of oilseed
	Safflower	*Carthamus tinctorius* L.	Compositeae	Drought tolerant
	Sunflower	*Helianthus annuus* L.	Compositeae	Photo-insensitive, Phototrophic movement
	Linseed (Flex)	*Linum usitatissimum* L.	Linaceae	Suitable for paira cropping. oilseeds and fiber crop
	Taramira	*Eruca sativa*	Cruciferae	Drought tolerant crop
	Castor	*Ricinus communis* L.	Euphorbiaceae	Drought tolerant crop
	Sesame	*Sesamum indicum* L.	Padaliaceae	Queen of oilseeds, Poorman's substitute for ghee
	Groundnut	*Arachis hypogea* L.	Fabaceae	Important oilseed of India
Fibre crops	Cotton	*Gossypium* spp.	Malvaceae	White gold
	Jute	*Corchorus* sp.	Tiliaceae	Green manuring and fiber crop
	Sunhemp	*Crotolaria juncea*	Fabaceae	Green manuring and fiber crop
Sugar crops	Sugarcane	*Saccharum officinarum* L.	Poaceae	Noble cane
	Sugarbeet	*Beta vulgaris* L.	Chinopodaceae	Substitute for sugarcane
Stimulant & medicinal crops	Tobbaco	*Nicotiana tabaccum*	Solanceae	Nicotine as a stimulant
		Nicotiana rustica	Solanceae	Nicotine as a stimulant
	Opium	*Papaver somniferum*	Papaveraceae	Green gold, addictive plant
Tuber crops	Potato	*Solanum tuberosum* L.	Solanceae	Poor man's friend
Fodder crops	Oat	*Avena sativa* L.	Poaceae	Prolamins (glute-min), named avenins
	Lucerne	*Medicago sativa* L.	Fabaceae	Queen of fodder
	Berseem/alfalfa	*Trifolium alexandrium* L.	Fabaceae	King of fodder

Maize | Sorghum | Pearlmillet

Clusterbean | Castor | Sesame

Blackgram | Greengram | Mothbean

Plate 1. Photographs of Crops

Paddy

Groundnut

Soyabean

Cotton

Pigeonpea

Cowpea

Wheat

Barley

Sugarcane

Plate 2. Photographs of Crops

Mustard

Linseed

Chickpea

Fenugreek

Isabgol

Cumin

Lucerne

Oat

Tobaco

Plate 3. Photographs of Crops

2

NUMERICAL EXERCISES ON PLANT POPULATION

Introduction

Plant population is defined as number of plants per unit area and it depends on the geometry or arrangement of crop plants in field. The main objective for maintaining an optimum plant population is to utilize the available resources for higher crop yields. Below or higher the optimum plant population causes less utilization of resources, higher weed infestation and more competition among crop plants results in lower yield. It is not possible to recommend a generalized plant population for particular crop as it varies considerably for different season, purpose of sowing and farming situations under different management practices.

Crop geometry

Crop geometry refers to the shape of the space available for individual plants. It influences crop yield through its influence on light interception, rooting pattern and moisture extraction pattern. Crop geometry is altered by changing inter and intra-row spacing (Planting pattern).

Plant density

Plant density is the number of plants per unit area in a cropped field. It indicates the size of the area available for individual plant.

Generalized formula for calculating plant population

$$\text{Plant population} = \frac{\text{Area (m}^2\text{)} \times 100 \times 100}{\text{Row to row spacing (cm)} \times \text{Plant to plant spacing (cm)}}$$

Example 1. Calculate the plant population of pigeonpea in one ha, if seed were sown at a spacing of 60 cm row to row and 25 cm plant to plant.

$$\text{Plant population (no. of plant ha)} = \frac{10000 \times 100 \times 100}{60 \times 20} = \frac{100000000}{1200} = 83{,}333 \text{ plants/ha}$$

Example 2. Calculate the plant population of maize for 2000 sq.m if seeds were sown at a spacing of 60 cm and 20 cm, row to row and plant to plant, respectively.

$$\text{Plant population} = \frac{2000}{60 \times 20} \times 100 \times 100 = 16{,}666 \text{ plants}$$

Table 1. Calculate plant population and complete the table for different major crops

Local Name	Scientific Name	Spacing (cm)		Plant population per ha
		Row to row	Plant to plant	
Rice	*Oryza sativa* L.	20	10	5,00,000
Maize	*Zea mays* L.	60	25	66,666
Sorghum	*Sorhum bicolor*	45	10	2,22,222
Pearlmillet	*Pennisetum typhoides*	45	15	1,48,148
Greengram	*Vigna radiate* L.	30	10	3,33,333
Blackgram	*Vigna mungo* L.	30	10	3,33,333
Moth	*Vigna acontifolia* L.	40	10	2,50,000
Cowpea	*Vigna sinensis* L.	45	10	2,22,222
Pigeonpea	*Cajanus cajan* L.	60	20	83,333
Horsegram	*Macrotiloma uniflorum* L.	30	10	3,33,333
Soybean	*Glycine max* L.	45	5	4,44,444
Clusterbean	*Cyamopsis tetragonoloba* L.	30	10	3,33,333
Castor	*Ricinus communis* L.	90	45	24,691
Sesame	*Sesamum indicum* L.	45	15	1,48,148
Groundnut	*Arachis hypogea* L.	30	15	2,22,222
		45	20	1,11,111
Cotton	*Gossypium* spp.			
	American	60	45	37,037
	Deshi	60	30	55,555
	Bt	100	100	10,000

Example 3. Calculate the plant population per hectare, if a plant occupy 0.045 square meter area.

$$\text{Plant population (No. of plants/ha)} = \frac{10000 \text{ m}^2}{0.045 \text{ m}^2} = 2{,}22{,}222$$

■■■■

3

SEED RATE OF DIFFERENT CROPS

INTRODUCTION

Seed rate is the quantity of seed required for sowing or for planting in unit area to obtain the desired plant population. The seed rate for a particular crop depends on its seed size (test weight/seed index - Table 1) and values of germination and purity percentage of particular seed lot. It is calculated on the basis of number of plants required per unit area, the germination and purity percentage of the seed lot under consideration.

Table 1. Test weight of major field crops

Name of crop	Test wt(g)	Name of crop	Test wt (g)
Rice	20-30	Sorghum	25-30
Wheat	36-50	Pearl millet	5-9
Barley	35-40	Chickpea (desi)	150-160
Maize	200-250	Lentil	17-38
Black gram	36-49	Greengram	20-70
Peas	120-220	Soybean	100-105
Groundnut	200-250	Rapeseed and mustard	3-5
Sunflower	40-50	Safflower	50-60
Castor	100-150	Cotton	70-100
Pigeonpea	110-120	Cowpea	180-210
		Tobacco	0.9-1.0

Factors affecting seed rate

Seed rate for a particular crop or variety is decided by the followings:

(a) Size and weight of seed (Higher seed rate for bold seeds, less seed rate for smaller seeds).

(b) Method of sowing (Higher seed rate for broadcasting than drilling and dibbling).

(c) Spacing [More the spacing (either within rows or between rows), less the seed rate and vice versa].

(d) Cropping system (Higher seed rate for pure crop than intercrops).

(e) Purpose of sowing (Grain, seed production, fodder, green manuring, intercropping etc).

(f) Farming situation (Higher seed rate in drylands than irrigated).

Calculation

A generalized formula for calculating seed rate in kg ha^{-1} is as under

$$\text{Seed rate (kg ha}^{-1}) = \frac{\text{Plant population in study area} \times 10000 \times \text{Test Weight}}{\text{Area under study (m}^2) \times 1000 \times 1000}$$

Or

$$\text{Seed rate (kg ha}^{-1}) = \frac{\text{No. of seeds sown in study area} \times 10000 \times \text{Wt. of one seed}}{\text{Area under study (m}^2) \times 1000}$$

If the values of germination percent and purity percent for a seed lot has been given then,

$$\text{Seed rate (Kg ha}^{-1}) = \frac{\text{Area (m}^2) \times \text{Test weight } \times 100 \times 100}{\text{Row to row spacing (cm)} \times \text{Plant to plant spacing (cm)} \times \text{Germination (\%)} \times \text{Purity (\%)} \times 1000 \times 1}$$

Or

$$\text{Seed rate (Kg ha}^{-1}) = \frac{\text{Area (m}^2) \times \text{Test weight} \times 100 \times 100}{\text{Spacing (m}^2) \times \text{Germination (\%)} \times \text{Purity (\%)} \times 1000 \times 1000}$$

Or

$$\text{Seed rate (kg ha}^{-1}) = \frac{\text{Area (m}^2) \times \text{Test weight} \times 100}{\text{Spacing (m}^2) \times \text{Real value} \times 1000 \times 1000}$$

Example 1

Calculate the seed rate of greengram (kg ha^{-1}) from the following details:

(a) Spacing –30 cm (row to row) × 10 cm (plant to plant)

(b) Germination % = 92

(c) Purity % = 85

(d) Test weight = 36 g

Solution

$$\text{Seed rate (kg ha}^{-1}) = \frac{\text{Area (m}^2) \times \text{Test weight} \times 100 \times 100}{\text{Spacing (m}^2) \times \text{Germination (\%)} \times \text{Purity (\%)} \times 1000 \times 1000}$$

By placing the values of different characters

$$\text{Seed rate (kg ha}^{-1}) = \frac{10000 \times 36 \times 100 \times 100}{0.30 \times 0.10 \times 92 \times 85 \times 1000 \times 1000}$$

$$\text{Seed rate (kg ha}^{-1}) = \frac{3600}{0.30 \times 0.10 \times 92 \times 85} = \frac{3600}{234.6} = 15.34 \text{ kg ha}^{-1}$$

Example 2

Calculate the seed rate of barley, if crop is sown at spacing of 20 cm row to row and 5.0 cm plant to plant and test weight of seed is 42 gms.

Solution

$$\text{Seed rate (kg ha}^{-1}) = \frac{\text{Area (m}^2) \times \text{Test weight} \times 100 \times 100}{\text{Spacing (cm}^2) \times 1000 \times 1000}$$

By placing the values of different characters

$$\text{Seed rate (kg ha}^{-1}) = \frac{10000 \times 42 \times 100 \times 100}{20 \text{ cm} \times 5 \text{ cm} \times 1000 \times 1000}$$

$$\text{Seed rate (kg ha}^{-1}) = \frac{4200}{20 \times 5} = 42 \text{ kg ha}^{-1}$$

Example 3

Calculate the seed requirement of maize for 6000 m^2 if sown at a spacing of 60 cm row to row and 25 cm plant to plant, respectively with a test weight of 240 gms.

Solution

$$\text{Seed rate (kg ha}^{-1}) = \frac{\text{Area (m}^2) \times \text{Test weight} \times 100 \times 100}{\text{Spacing (cm}^2) \times 100 \times 100}$$

By placing the values of different characters

$$\text{Seed rate (kg ha}^{-1}) = \frac{10000 \times 240 \times 100 \times 100}{60 \times 25 \times 1000 \times 1000}$$

$$\text{Seed rate (kg ha}^{-1}) = \frac{2400}{150} = 16 \text{ kg ha}^{-1}$$

Seed requirement for 6000 m^2 (kg)

$$= \text{Seed rate (kg ha}^{-1}) \times \frac{\text{Area to be sown (m}^2)}{10000}$$

$$= 16 \times \frac{6000}{10000} = 9.6 \text{ kg}$$

Example 4

Calculate seed rate of upland rice on the basis of following data, if 12% seeds damaged

(a) Spacing – 20 cm x 3 cm

(b) Germination % - 89

(c) Purity % = 82

(d) Test weight – 26 g

Solution

Total number of seeds required = No. of seeds required/ha + No. of seeds needed for replacing damaged seeds

$$\text{No. of seeds required/ha} = \frac{10000}{\text{Spacings}} = \frac{10000}{.20 \times .03} = 1666666.6 \text{ seeds}$$

No. of damaged seeds/ha = 10000 x 12 =120000

Therefore total no. of seeds = 1666666.6 + 120000 = 1786666.6 seeds/ha

Thus, seed rate would be (kg ha)

$$= \frac{\text{Total no. of seeds / ha} \times \text{test wt.} \times 100 \times 100}{1000 \times \text{germination \%} \times \text{purity \%} \times 1000}$$

$$= \frac{1786666.6 \times 26 \times 100 \times 100}{1000 \times 89 \times 82 \times 1000} = 63.65 \text{ kg ha}^{-1}$$

Example 6

Calculate seed rate of groundnut in terms of kernels as well as unshelled pods:

(a) Spacing 50 cm x 15 cm

(b) Germination % = 90

(c) Purity % = 84

(d) Test weight of kernels - 650 g

(e) Shelling % – 75

Solution

$$\text{Seed rate of kernels (kg ha–1)} = \frac{10000 \times 650 \times 100 \times 100}{0.5 \times 0.15 \times 90 \times 84 \times 1000 \times 1000} = 114.6$$

Quantity of unshelled pods needed for sowing in kg/ha

$$= \frac{\text{Amount of kernels in kg/ha} \times 100}{\text{Shelling per cent}} = \frac{114.6 \times 100}{75} = 191 \text{ kg ha}^{-1}$$

Student Exercise

Calculate the seed rate of important cereals, pulses, oil seeds, vegetables, tuber crops, fodder crops, green manure crops, etc. Collect the information for required parameters and note down in following table using recommended spacing from any text book of field crops or using values given in previous exercise.

Table 2. Calculation of seed requirement for *kharif* crops in kg ha^{-1}

Local Name	Scientific name	Test weight (g)	Germination percentage	Purity percentage	Seed rate (Kg ha^{-1})
Maize	*Zea mays* L.				
Sorghum	*Sorhum bicolor*				
Pearlmillet	*Pennisetum typhoides*				
Greengram	*Vigna radiata* L.				
Blackgram	*Vigna mungo* L.				
Moth	*Vigna acontifolia* L.				
Cowpea	*Vigna sinensis* L.				
Arhar	*Cajanus cajan* L.				
Horsegram	*Macrotiloma uniflorum* L.				
Soybean	*Glycine max* L.				
Clusterbean	*Cyamopsis tetragonoloba* L.				
Castor	*Ricinus communis* L.				
Sesame	*Sesamum indicum* L.				
Groundnut	*Arachis hypogea* L.				
Cotton	*Gossypium* sp.				
Jute	*Corchorus* sp.				
Sunhemp	*Crotolaria juncea*				

4

CALCULATION OF SEED PURITY PERCENTAGE

INTRODUCTION

The seed rate is a inversely proportional to germination and purity percentage of the seed lot under testing. The knowledge about purity and germination test values is the prime indicators for selecting or rejecting a seed lot used for sowing. The main objective to conduct purity test is to get an accurate and reproducible results related to planting value of the seed lot.

Purity percentage is defined as ratio of weight of pure and viable seed to the weight of seed sample or lot. Its calculation is simple one in which different components viz., pure seed, other crop seed, weed seeds and inert matter are weighed separately and then percentage of pure seed fraction to total seed lot is called as purity % as grain below. Germination percentage is calculated using these pure seeds.

Percent of each component or fraction to the extent of 5% or more should be calculated and reported while components of less than 0.05% should be reported as 'traces'. The purity percentage is expressed as:

$$\text{Purity (\%)} = \frac{\text{Weight of pure seed}}{\text{Weight of sample seed}} \times 100$$

$$\text{Percentage of individual fraction} = \frac{\text{Weight of individual fraction}}{\text{Weight of sample seed}} \times 100$$

Seed lots possessing less than minimum limits of purity must be rejected because these seeds will result in:

- Poor plant population and ultimate the lower yield.
- Poor price of produce in market
- Higher cost of cultivation.
- Higher cost of seed itself.
- Spread of new diseases, insect pests and noxious weeds.

Calculation

Calculate the purity percentage of a seed sample weighing 220 grams using following observation:

Components/fractions		Weight	Percentage of each fraction
a	Other crop seeds	19	8.6
b	Weed seeds	3	1.4
c	Inert matter	11	5.0
d	Pure seed {Total weight-(a+b+c)}	200-33=187	85.0
	Total	220	100.0

Solution

$$\text{Purity (\%)} = \frac{\text{Weight of pure seed}}{\text{Weight of sample seed}} \times 100 = \frac{187}{220} \times 100 = 85\%$$

■■■■

5

CALCULATION OF GERMINATION PERCENTAGE

INTRODUCTION

Germination is the emergence and development of a seedling from seed embryo or those essential structures which for the seed being tested, indicates its ability to develop into a normal plant under favorable conditions in the field/soil. It is essential to conduct the germination test before sowing to obtain accurate and reproducible results relating to planting value of the seed lot. Germination tests are conducted on composite samples, obtained by mixing primary samples collected at random from the stored seed lot. A suitable size of seed sample is taken and favorable conditions are provided for germination. As per ISTA (International Seed Testing Association) rules, a minimum of 400 seeds while as per national rule 200 seeds are required depending on the seed size. Germination percentage is calculated by counting healthy and germinated seedlings to the number of seeds sown after passing suitable period prescribed for different crop. Based on the size of seeds and substrate for germination, different methods are used for testing the germination. The germination of seeds depends on viability of seeds, dormancy and environmental conditions. Germination test is conducted with the seeds from pure seed fraction after purity test.

METHODS

1. **Petri Dish Method:** Both TP (Top of Paper) and BP(Between Paper) method can be conducted in the Petri dish. Two filter papers or blotters are placed on the bottom of a petri dish (approximately 8 cm diameter) and are soaked with water. Convenient number of seeds i.e., 10-20 numbers depending on seed size are placed on the surface of paper in the petri dish. The kind of seed, date, and time of seed soaking should be written on glass cover. Germination percentage is calculated on the basis of 100-200 seeds. It is suitable for small seeds - tomato, A*maranthus*, chilli, etc.

2. **Rolled Towel Test:** Two wet towels are placed on a smooth table top, appropriate number of seeds are placed on the upper side of towel and covered by another wet towel. A fold is made on bottom of towel to prevent the seed from falling out. These towels are rolled from right to left. The kind of seed, date, time of sowing are noted on the towel fold.
3. **Rolled Paper Towel Method:** Two wet paper towels specially made for germination test are placed on the working table. Surface of the towel is marked into two halves. The right half of the towel is placed with seeds and left half is used to fold the right half. Details regarding date of soaking and time of sowing are written on fold. Water is sprinkled frequently on towel. Observations are taken periodically. The towels are kept on rack in single layer. This method is suitable for larger seeds
4. **Sand Method:** Aluminium pans 23 cm × 23 cm × 8 cm size are taken as containers. Fine river sand is spread on them. Top surface is moistened with water. Seeds are placed on the surface and then covered with thin layer of sand of 1.25-2.25 cm thickness. The pans are labelled with name of seed, planting date and time of planting. Germination count is taken periodically. This method gives lower values of germination when compared to other three methods. But data are comparable to field conditions.
5. **Rag Doll Method**: Rags or gunny bags are used in this method. The seeds to be tested are arranged in moistened gunny bags which are rolled and tied loosely at the bottom of the bundle which is also kept on a rack for germination. Rolls are opened after a fixed period of time.
6. **Germination Test through Germinators**: Equipments specially designed to test germination are called seed germinators. Seed may be placed in petri dish/sand/blotting paper and water can be supplied using sprayer. In sophisticated germinators, moisture, humidity, temperature can be maintained. These incubators are electrically operated with provision for water.

Calculations:

$$\text{Germination (\%)} = \frac{\text{Number of seeds germinated}}{\text{Total number of seed sown}} \times 100$$

Table 1. Calculate the germination percentage of different crops using following observations.

Crop	Number of seeds sown	Number of seeds germinated
Maize	44	38
Cowpea	52	42
Wheat	60	52
Mustard	120	100

Solution

$$\text{Germination (\%) for maize} = \frac{\text{Number of seeds germinated}}{\text{Total number of seeds sown}} \times 100$$

$$= \frac{38}{44} \times 100 = 86.4\%$$

Similarly for cowpea

$$= \frac{42}{52} \times 100 = 80.8\%$$

For wheat

$$= \frac{52}{60} \times 100 = 86.7\%$$

For mustard

$$= \frac{100}{120} \times 100 = 83.3\%$$

Note: According to ' Indian' Seed Act 1966 ' which has been in force from 1st October, 1969, seeds must have minimum prescribed germination percentage. Different categories of seed viz., Nucleus, Breeder, Foundation, Certified or TFL seed have different standardized values of purity and germination percentage. Similarly, the different crops have different values of these parameters. Below the standard, the seed lot under study may be rejected.

Real Value of Seed

It represents the seed quality in terms of purity and germination. It is function of purity and germination percentage and expressed in percentage. This is the standard measure for comparison of different seed lots having different values of purity and germination percentage. A seed sample having higher real value is always better than lower one as explained in next page:

Sample A

Purity % 85

Germination % 95

Sample B

Purity % 98

Germination % 80

Real value for 'A' $= \frac{85 \times 95}{100} = 80.75\%$ 'B' $= \frac{98 \times 80}{100} = 78.40\%$

It is more economical and better to buy the seed from seed lot of sample 'A'. A seed lot having < 70 real value is not preferred for sowing purpose because of poor germination and purity values.

6

TESTS FOR ASSESSING SEED VIABILITY

INTRODUCTION

Seed tested for germination is the most reliable and satisfactory criteria for choosing a seed lot for sowing. But the germination methods are slow and time consuming. Hence, it is required to conduct a rapid test to find out whether the seeds are viable or not. The viability is defined as capacity of the seed to germinate or capacity of seed to remain capable of germination for some specific period of time. The viability of seed is affected by moisture content, storage conditions, age of the seed and its genetic constitution.

METHODS OF ASSESSING VIABILITY TEST

Electrical conductance method

Seeds are soaked in distilled water for 24-48 hours and electrical conductance of the solution is tested. As the seeds deteriorate, substances (metabolites) can be leached through the cell membrane which becomes permeable due to degradative processes in proportion to the degree of deterioration. Hence, measurement of EC can be correlated with seed quality. However, the relationship of conductivity to germination must be established for each species.

Permanganate method

It is the quantitative test to find out whether the seeds are viable or not. Seeds are soaked in $KMnO_4$ solution. The dead tissues develop discolouration due to the release of metabolites.

Aniline dye method

Seeds are soaked in indigo carbine for few hours. Dead seeds become stained and the viable seeds can be counted and percentage viability can be worked out from this.

Embryo culture method

Embryo is carefully removed from the cotyledon of endosperm and placed in granulated peat moss or in a sterilized nutrient agar medium. This method takes about 7-10 days to give viability %. The viability is judged by counting the number of germinated embryos.

Tetrazolium chloride test

ISSTA (Indian Society of Seed Testing Associations) recommended this method for testing the viability for seeds of a number of tree species. In this method, a colourless solution of tetrazolium (2,3,5- triphenyl tetrazolium chloride), an oxidation reduction indicator is used. When treating the seeds with tetrazolium, it interferes with the respiratory process and accepts H^+ ion from hydrogenase. By hydrogenation it will form a red stable substance called triphenyl formasan in the living cells. This makes it possible to distinguish between dead seeds and viable seeds.

Procedure

- Seeds should be soaked in water overnight at room temperature.
- The water soaked seeds are then cut longitudinally (monocotyledons) to expose the embryo or seed coats of dicots should be removed to facilitate the quick penetration of tetrazolium.
- The desired number of seeds should be soaked in 1% TZ and kept in dark for 3-4 hours at temperature of 30ºC.
- When the color has developed, TZ should be drained and seed should be rinsed 2-3 times with water and evaluated. The viable or living seeds turn bright red colouration with intensive colouration in the embryo while the dead seeds remain in their original color.

$$\text{Viability (\%)} = \frac{\text{Number of seeds strained}}{\text{Total number of seeds tested}} \times 100$$

Calculation

Calculate the viability percentage of different crops using following observations:

Crop	Number of seeds sown	Number of seeds strained (coloured)	Viability (%)
Maize	44	38	86.4
Cowpea	52	42	80.8
Wheat	60	52	86.7
Mustard	120	100	83.3

■■■■

7

IDENTIFICATION OF MANURES AND FERTILIZERS

Introduction

The chemical elements necessary for plant growth are termed as plant nutrients and the process of absorption, translocation and assimilation of these nutrients by the plants is known as plant nutrition. As per the criteria of essentiality, total 17 nutrients have been considered as essential plant nutrients (Table 1). The sources or materials supplying these nutrients are termed as manures and fertilizers. The knowledge of different kinds such as organic manures (either bulky or concentrate, green manures and green leaf manures), chemical and bio-fertilizers and their nutrient contents is very important for increasing the fertilizer use efficiency balanced nutrition and higher production along with improvement in soil fertility (Annexure 13).

Table 1. Essential nutrients and their forms of absorption

Absorbed as single nutrient ion		Absorbed in a combined form	
Nutrient elements	**Forms absorbed by plants**	**Nutrient elements**	**Forms absorbed by plants**
Potassium	K^+	Nitrogen	Ammonium (NH_4^+) and Nitrate (NO_3^-)
Calcium	Ca^{++}	Phosphorus	$H_2PO_4^-$, $HPO_4^=$
Magnesium	Mg^{++}	Molybdenum	$MoO_4^=$ (Molybdate)
Iron	Fe^{++} (Ferrous), Fe^{+++} (Ferric)	Sulphur	$SO_4^=$
Manganese	Mn^{++} (Manganous), Mn^{++++} (Manganic)	Boron	H_3BO_3 (Boric acid), $H_2BO_3^-$ (Borate), HBO_3^{2-}, BO_3^{3-}
Copper	Cu^{++}	Hydrogen	H_2O, H_+
Zinc	Zn^{++}	Carbon	CO_2.

Absorbed as single nutrient ion		Absorbed in a combined form	
Nutrient elements	**Forms absorbed by plants**	**Nutrient elements**	**Forms absorbed by plants**
Chlorine	Cl^{++}	Oxygen	CO_2, O^{2-}, OH^-
Silicon	Si^{++++}, $Si(OH)_4$		
Cobalt	CO^{2+}		
Sodium	Na^+		
Nickel	Ni^{++}		

Manures

Manures are organic materials of plant or animal origin incorporated into soil well before sowing/planting to give sufficient time for their decomposition and release of nutrients. Organic manures can either be bulky or concentrated according to its nutrient content.

Bulky organic manures: Bulky organic manures contain small percentage of nutrients and are applied in large quantities (bulk) 2 to 4 weeks before sowing e.g. farmyard manure (FYM), compost, green-manure etc. They supply almost all essential plant nutrients including micronutrients and improve soil properties like soil structure, water holding capacity and organic matter (humus) content in soil.

Concentrated organic manures: Concentrated organic manures have higher percentage of plant nutrient as compared to bulky organic manure. They are usually raw material or by products of animals and plants origin and are applied at the time of field preparation as basal application. The important concentrated organic manures are as follows:

Oilcakes:

1. Edible oil cake : Can be safely fed to livestock

   eg: Coconut, Groundnut, Cotton, Sesamum etc

2. Non edible oilcakes : Some anti-nutritional substances limit their feedling to animals (Annexure 9)

   eg: Castor, Neem, Mahua, Pongamia etc

Animal meal

- Blood meal
- Bone meal
- Hoof meal
- Fish meal

Table 2. Nitrogen, Phosphorus and Potassium content of various bulky and concentrates organic manures

Manure	N %	P_2O_5 %	K_2O %
Bulky organic manures			
Cattle dung (Fresh)	0.4-0.5	0.3-0.4	0.3-0.4
Cattle dung (Dry)	5.0	1.4	3.8
Sheep manure	3.0	1.0	2.0
Fresh poultry manure	1.0-1.8	1.4-1.8	0.8-0.9
Concentrate organic manure			
Bone meal (Raw)	3.0-4.0	20.0-25.0	-
Bone meal (Steamed)	1.0-2.0	25.0-30.0	-
Blood meal	10.0-12.0	1.0-2.0	1.0
Meat meal	10.5	2.5	0.5
Horn and Hoof meal	13	-	-
Fish meal	4.0-10.0	3.0-9.0	0.3-1.5
Compost (Farm)	0.5	0.26	0.5
Compost (Town)	1.5-2.0	1.0	1.5
Vermicompost	1-1.8	0.1-1.9	0.5-1.6
Farm Yard Manure	0.5	0.25	0.5
Coir pith compost	1.20	1.20	1.20
Press mud	1.0-1.5	4.0-5.0	2.0-7.0
Non-edible oil cakes			
Castor oil cake	4.3	1.8	1.3
Mahua cake	2.5	0.8	1.8
Neem cake	5.2	1.0	1.4
Safflower cake	4.9	1.4	1.2
Cotton cake	3.9	1.8	1.6
Karanj cake	3.9	0.9	1.2
Edible oil cakes			
Coconut cake	3.0	1.9	1.8
Groundnut cake	7.3	1.5	1.3
Rape seed cake	5.2	1.8	1.2
Sesamum cake	6.2	2.0	1.2
Niger cake	4.7	1.8	1.3
Linseed cake	4.9	1.4	1.3

Source: (Organic manures by Dr. Krishan Chandra, 2005)

Green manures: Practice of ploughing under composed green plants into soil for purpose of improving soil fertility. Green manures are bulky organic manures. It is of two types.

- **Green manuring:** Green manure crops are grown and buried in the same field. Crop should be quick growing crop, often a legume, and plough it *in situ* at flowering stage.

Fig. 1. *In-situ* green manuring.

- **Green leaf manuring***:* Collection and transportation of green leaves and twigs of trees, shrubs and herbs from elsewhere and incorporation in soil through ploughing is known as green leaf manuring.

Table 3. Green manures, green leaf manures and their composition

Crop Trees	Botanical name	N %	P_2O_5 %	K_2O %
	Green manure			
Sunnhemp	*Crotalaria juncea*	2.30	0.50	1.30
Sesbania	*Sesbania rostrata*	3.30	0.60	1.20
Sesbania	*Sesbania aculeata*	3.20	0.60	1.20
Pillispera	*Phaseolus trilobus*	2.80	0.50	1.15
Sesbania	*Sesbania speciosa*	2.71	0.53	2.21
Kalinji	*Tephrosia purpurea*	3.10	0.52	1.18
	Green leaf manure			
Glyricida	*Glyricida sepium*	2.76	0.28	4.60
Pongamia	*Pongamia glabra*	3.31	0.44	2.39
Neem	*Azadirachta indica*	2.83	0.28	0.35
Gulmohar	*Delonix regia*	2.76	0.46	0.50
Peltophorum	*Peltophorum ferrugenum*	2.63	0.37	0.50
Cassia	*Cassia fistula*	1.60	0.24	1.20
Subabool	*Leucaena leucocephala*			

Other examples of green leaf manures

- *Calotropis gigantia*
- *Pongamia pinnata*
- *Tephrosia* sp
- *Cassia tora*
- *Macaranga peltata*
- *Terminalia paniculata*

Fertilizers: Fertilizers are mined natural deposits or industrially manufactured materials containing one or more essential plant nutrients relatively in large quantities. The term fertilizer is essentially used for materials supplying major plant nutrients (i.e., N, P and K).

NITROGENOUS FERTILIZERS

Nitrogenous fertilizers are applied to supply nitrogen in different forms.

Based on forms of nitrogen they are classified as:

- Ammonical fertilizers
- Nitrate fertilizers
- Ammonical nitrate fertilizers
- Amide fertilizers

1. Ammonical fertilizers : The nitrogen is present in ammonia form eg. Ammonium sulphate, Anhydrous ammonia, Ammonium phosphate.

2. Nitrate fertilizers: Nitrogen is in nitrate form eg. $NaNO_3$, KNO_3. They are highly soluble in water and highly mobile.

3. Ammonical and nitrate fertilizers: Nitrogen is present both as nitrate and ammonia eg. NH_4NO_3, $(NH_4)_2SO_4$. NH_4NO_3 (ASN) and CAN (Calcium Ammonium Nitrate).

4. Amide fertilizers: Nitrogen is present in the amide form eg. Urea $[CO(NH_2)_2]$.

Table 4. Nitrogenous fertilizers and their composition

Fertilizer	Chemical formula	N %	Other nutrients
Ammoniacal fertilizer			
Ammonium Sulphate	$(NH_4)_2SO_4$	20.6	24.0 % S
Ammonium chloride	NH_4Cl	25.0	
Anhydrous ammonia	NH_3	81.5	

Fertilizer	Chemical formula	N %	Other nutrients
Liquor ammonia	NH_3 in water	20.0	
Nitrate fertilizer			
Sodium Nitrate	$NaNO_3$	16.0	
Calcium Nitrate	$Ca(NO_3)_2$	15.5	
Potassium Nitrate	KNO_3	13.8	
Ammonical and Nitrate fertilizer			
Ammonium Nitrate	NH_4NO_3	33.5	
Calcium Ammonium Nitrate	$NH_4NO_3.CaCO_3$	26.0	
Ammonium Sulphate Nitrate	$(NH_4)_2SO_4.\ NH_4NO_3$	26.0	15.0 % S
Amide fertilizer			
Urea	$CO(NH_2)_2$	46.0	
Calcium Cynamide	$CaCN_2$	20.6	

PHOSPHATIC FERTILIZERS

Based on solubility, phosphorus fertilizers are classified as follows:

(1) Water soluble forms: These fertilizers contain phosphorus in available form which can be readily absorbed by plants in neutral soils. In acidic and free iron and/or aluminium containing soils, the soluble P will convert into hydroxy phosphates and in alkaline and calcareous soils, water soluble phosphorous is converted into insoluble calcium phosphate. They are normally recommended for short duration field crops. Single super phosphate, double super phosphate, triple super phosphate and ammonium phosphate are water soluble phosphatic fertilizers.

(2) Citric acid soluble forms: The citric soluble fertilizers are suitable for acidic soils because at low pH, they react with citric and other organic acids and converted into monocalcium phosphate. Dicalcium phosphates like basic slag and calcium meta phosphate are citric acid soluble. They are applied to long duration crops like sugar cane, tapioca, red gram, etc.

(3) Citric acid insoluble forms: These mineral fertilizers contain phosphorus, which is insoluble in water as well as in citric acid. They are suitable in strongly acid soils or organic soils. They are suitable for plantation crops like tea, rubber, coffee, etc. These fertilizers are given in green manured fields. The phosphorus is very slowly released by microbes at action and remains in soil for long time. Tricalcium phosphates like bone meal, steamed bone meal, rock phosphate, etc are citric acid insoluble fertilizers.

Table 5. Phosphatic fertilizers and their composition

Fertilizer	Chemical formula	% of P_2O_5	Other nutrients
Water soluble phosphatic fertilizers			
Single superphosphate	$Ca(H_2PO_4)_2.\ CaSO_4$	16-18	12.0% S 21% Ca
Double superphosphate	$CaH_4(PO_4)_2$	32	
Triple superphosphate	$Ca(H_3PO_4)_2$	46 to 48	
Ammonium phosphate		20	
Citric acid soluble phosphatic fertilizers			
Dicalcium phosphate	$Ca_2(HPO_4)_2$	14	
Basic slag	$(CaO)5P_2O_5.SiO_2$	20 (15-18)	
Calcium metaphosphate	$Ca\ (PO_3)_2$	60 to 64	
Water and citrate insoluble phosphatic fertilizers			
Rock phosphate	$[Ca_3(PO_4)_2]_3$	20 to 38	
Bone meal	$(Ca(PO_4)\ _2)_3\ CaF_2$	21 to 25	

POTASSIC FERTILIZERS

Potassium fertilizers are obtained from deposits of several feet below the earth's surface and from the brines of lakes and seas. All potash salts used as fertilizers are water soluble. In India, most of K-fertilizer are imported from other country.

Table 6. Potassic fertilizers and their composition

Sl.No.	Fertilizer	Chemical formula	% of K_2O	Other nutrients
1	Muriate of potash	KCl	60	
2	Sulphate of potash	K_2SO_4	48-50	18.0% S
3	Potassium nitrate	KNO_3	44	32% N
4	Potassium magnesium sulphate	$K_2SO_4.\ 2MgSO_4$	25-30	

Complex fertilizer/Compound fertilizer : A fertilizer intended to supply more than one major plant nutrient. The production of these involves some chemical reaction. These are either in ratio or content.

Table 7. Complex fertilizers and their composition

Sl. No.	Fertilizer	% of N	% of P_2O_5
1	Di-Ammonium phosphate	18	**46**
2	Ammonium phosphate sulphate	16	**20**
3	Urea ammonium phosphate	28	**28**

Micronutrient supplying fertilizers: Plant needs 8 essential micro nutrients. Their quantity is usually required in ppm.

Table 8. Micronutrient supplying fertilizers and their contents

Fertilizer	% of Nutrient	Fertilizer	% of Nutrient
Chelated Zn as Zn EDTA	12% Zn	Ammouinium molybdate	52% Mo
Chelated Fe as Fe EDTA	12% Fe	Borax $(NH_4)_6.MO_7O_{24}.4H_2O$	19% B
Zinc sulphate $(ZnSO_4.7H_2O)$	21% Zn, 15% S	Elemental sulpher	100% S
Ferrous sulphate $(FeSO_4.7H_2O)$	19% Fe, 19% S	Gypsum	20% S
Copper sulphate $(CuSO_4.5H_2O)$	24% Cu, 13% S	Maganes sulphate	30.5% Mn, and 17% S

Biofertilizers

Biofertilizers are defined as preparations containing living cells or latent cells of efficient strains of microorganisms that help crop plants' to uptake of nutrients by their interactions in the rhizosphere when applied through seed or soil treatment. They accelerate certain microbial processes in the soil which augment the extent of availability of nutrients in a form easily assimilated by plants.

Biofertilizers add nutrients through the natural processes of fixing atmospheric nitrogen, solubilizing phosphorus and stimulating plant growth through the synthesis of growth promoting substances. Biofertilizers particularly *Rhizobium* are strain specific and described in Annexure 10.

Table 9. Groups of bio-fertilizers and their examples

S. No.	Groups	Examples
N fixing Biofertilizers		
1.	Free-living	*Azotobacter, Beijerinkia, Clostridium, Klebsiella, Anabaena, Nostoc*
2.	Symbiotic	*Rhizobium, Frankia, Anabaena azollae*
3.	Associative Symbiotic	*Azospirillum*
P Solubilizing Biofertilizers		
1.	Bacteria	*Bacillus megaterium* var. *phosphaticum, Bacillus subtilis, Bacillus circulans, Pseudomonas striata*

2.	Fungi	*Penicillium* sp., *Aspergillus awamori*
P Mobilizing Biofertilizers		
1.	Arbuscular mycorrhiza	*Glomus* sp., *Gigaspora sp.*, *Acaulospora* sp., *Scutellospora* sp. and *Sclerocystis* sp.
2.	Ectomycorrhiza	*Laccaria* sp., *Pisolithus* sp., *Boletus* sp., *Amanita* sp.
3.	Ericoid mycorrhizae	*Pezizella ericae*
4.	Orchid mycorrhiza	*Rhizoctonia solani*
Biofertilizers for micro nutrients		
1.	Silicate and Zinc solubilizers	*Bacillus* sp., *Saccharomyces* sp., *Penicillium simplicissimum*
Plant Growth promoting rhizobacteria		
1.	Pseudomonas	*Pseudomonas fluorescens*

Source: http://www.krishisewa.com/articles/organic-agriculture/115-biofertilizers.html.

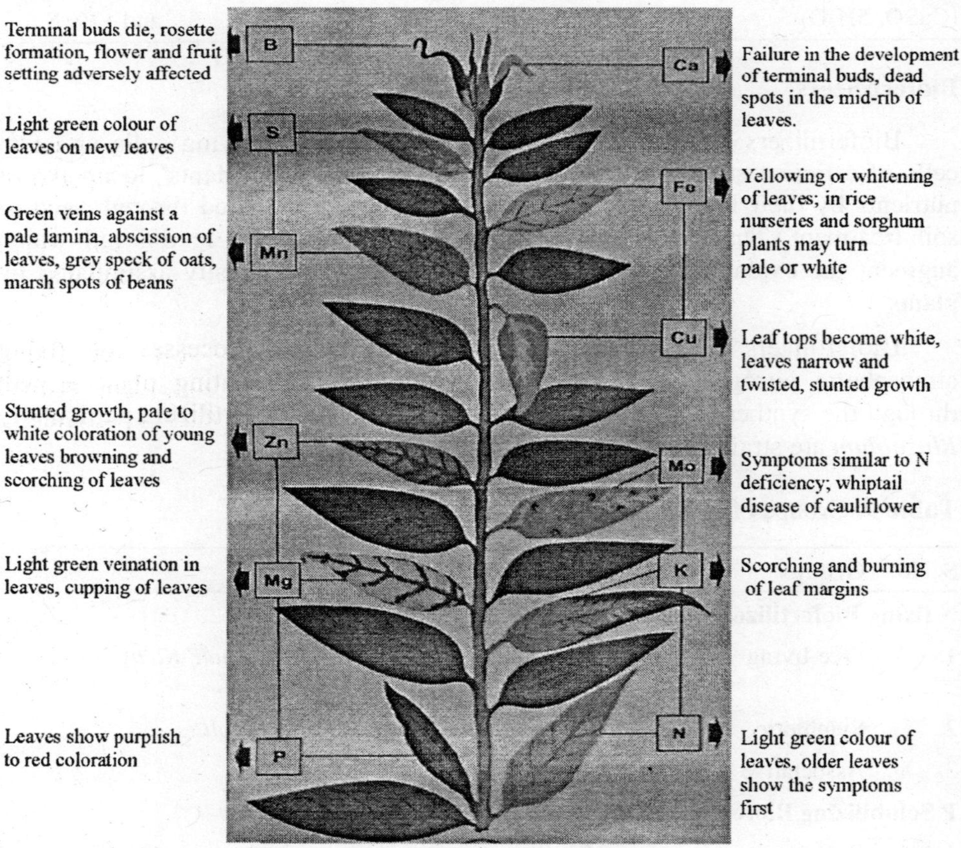

Fig. 2. Symptoms of deficiency of essential nutrients in different plants parts.

Plate 1. Photographs of manures and fertilizers

8

METHODS OF FERTILIZER APPLICATION

INTRODUCTION

As described in chapter 7, plant needs 17 essential elements to complete their life cycle. The sources of supply of these nutrients are called as manures and fertilizers. The fertilizers are applied to the soil and on foliage undergoes various reactions resultant into a considerable loss causing lower nutrient use efficiency. These losses does not influence the yield but also reduce the sustainability of soils. The knowledge regarding methods of fertilizer application is one of four principles of nutrient management i.e. right time, right dose, right sources and right methods of application to obtain highest nutrient recovery from these sources. The choice of method and time of fertilizer application depends on their kind and amount, nature of field crops and their nutrient requirements, soil properties, convenience in application, efficiency and safety in fertilizer application. Thus, it is very important to acquaint with different methods of application of fertilizers to achieve highest fertilizer use efficiency.

Principles involved in selecting methods of application

- Nitrogen fertilizers applied on the soil surface reach the plant roots easily and readily so these can be broadcasted before sowing.
- Since nitrogen fertilizers are liable to be lost by leaching, their application on soil surface followed by irrigation is enough to meet the requirement at the critical stage of crop growth. Hence, they are suitable for top dressing or side dressing.
- P should be localized placed (close to plant roots) where it will be readily accessible to plant roots.
- Localized placement is also practised for K fertilizers to avoid fixation losses in acid soils because it moves slowly in soil, hence should be applied near the *rhizosphere*.

Methods of fertilizer application: Following are the methods of fertilizer application:

I. Solid Form

(1) Broadcasting: The manures and fertilizers scattered uniformly over the entire field before or after sowing of the crop and are incorporated by tilling or cultivating is termed as broadcasting. This method proves effective when the crops have a dense stand, plant roots absorb nutrient from whole volume of soil, soil is rich in fertility, large amount of material is to be used etc. Though this method is easy, less time consuming, cheap and more convenient to the farmers, yet, it is not advantageous because it encourages weed growth, low recovery (as most of the material remains on the soil surface and does not reach upto the site of absorption i.e. root zone for uptake by plants, loss of fertilizer nutrients due to washing, run-off, volatilization, etc. Hence the recovery (extent of fertilizer used by plants) ranges between 25 and 45 per cent or even less. Broadcasting may be done in following ways:

(i) **Basal application:** Spreading of fertilizers before sowing or planting of the crops and mixing them by cultivating the soil during seed-bed preparation is termed as basal application.

(ii) **Top-dressing:** Spreading of fertilizer in standing crops is termed as top dressing without considering the crop rows but when the crop rows are taken into account and the material is dropped on the ground surface near the crop rows, then it is called as side dressing.

(2) Drilling or placement - It refers to application of fertilizers into the soil furrows formed at the desired depth where the crop roots can take them easily and in maximum amount while losses through uptake by weeds, washing, run-off, volatilization etc. could be eliminated to the greatest extent. Placement can be done by the following ways.

(i) ***Plough sole placement*** - Fertilizers are applied or dropped in the plough sole level, which will be covered by the plough during the opening of adjacent furrow. This could be done by kera, pora etc.

(ii) ***Deep placement*** - Fertilizers or manures are placed at the bottom of the top soil at a depth of 10-12 cm, especially in the puddled rice fields. This method is also adopted in dry land condition where the fertilizers are placed deeper than plough sole level. This helps in maximum root elongation and also eliminates various losses of nutrients from the soil.

(iii) ***Sub soil application*** - When fertilizers are placed still deeper than the seeding or planting depth (at a depth >15 cm) and also deeper than the previous two methods. This is especially followed for tree crops and orchard crops.

(3) **Localized or spot placement:** Fertilizers are placed in the root zone or near the roots from where roots can absorb easily. There is feature between placement and localized placement. The former refers to applying fertilizer into the soil without special reference to the location of seed or plant while the latter implies the application of fertilizer into the soil close to the seed or plant.

Localized placement of fertilizers has many advantages over broadcasting such as relatively less quantity is required, weed growth is suppressed, various losses are reduced, fertilizers are placed in moist zone to make it easy available to crop roots, higher fertilizer recovery, increased response of crops etc. to applied fertilizer. But the method is very technical and needs special precaution. It is very expensive also.

Generally placement of fertilizer is done for three reasons

- Efficient use of plant nutrients from plant emergence to maturity.
- To avoid the fixation of phosphate in acid soils.
- Convenience to the grower.

The method could be adopted in following ways

(*i*) ***Contact of drill placement*** - Fertilizers or manures are placed at the time of drilling or placing the seeds for making good contact of fertilizers or manures with seeds or seedlings. This is performed by seed-cum-fertilizer drill.

(*ii*) ***Band placement*** - This is the localized placement of manures or fertilizers at about 5 cm away from the seed or plant in any direction. This is a placement of fertilizers by the side of plants in the hill for widely spaced plants like maize (termed at discontinuous bands) or along the rows of the crop for closely spaced crops like cereals, minor millets (termed as continuous band placement). This method has a definite relationship of fertilizers with seedlings or seeds, hence, this method gives very promising results when soil surface is dry.

(*iii*) ***Pocket placement*** - Application of fertilizers deep in soil to increase its efficiency. It is especially followed in very widely spaced crops like sugarcane where fertilizers are put in 2 to 3 pockets open and around every hill by means of a sharp stick.

(*iv*) ***Side dressing*** - It refers to hill and ring placement of manures or fertilizers. It consists of spreading the fertilizer between the rows or around the plants.

(*v*) ***Pellet application*** - Nitrogen fertilizers are pelleted like mud ball or urea super granules (USG) and placed deep (10 cm) into the saturated soils (reduced zone) of wet land rice to avoid nitrogen loss from applied fertilizers. These balls or pellets are dropped near the crop rows in rice or jute under deep water conditions.

Methods of application for liquid fertilizers

Plants can absorb nutrients not only through roots but also through stomata. This property is used for spray application of fertilizers onto the foliage of crops. Use of liquid fertilizers is regularly increasing in India. It is most suitable method under dry land agriculture and in the areas which are prone to erosion problems. Liquid fertilizers may be applied in following ways

(i) **Use of starter solution:** They are solutions of fertilizers prepared in low concentrations and used for soaking seeds, dipping roots, spraying on seedlings etc. in nutrient deficient areas for early establishment and growth of crops particularly for transplanted crops. Starter solutions usually contain N, P, K in 1 : 2 : 1 or 1 : 1 : 2 ratio. Thus, it serves as irrigation water as well as nutrient solution for the crops.

(ii) **Fertilizer application through irrigation water or Fertigation:** It is the application of fertilizer dissolved in irrigation water in either open or closed system *i.e.*, lined or unlined open ditches and sprinkler or trickle systems, respectively.

(iii) **Nutrient injection method:** The anhydrous ammonia is injected into the soil at a depth of about 20-25 cm and at a pressure of about 200 (psi) pound per square inch with special injecting equipments. The anhydrous ammonia is the cheapest source of nitrogen because of its lower unit value. This methods is successful when the soil should have fine tilth, enough moisture so the loss of nitrogen in the form of ammonia does not take place.

(iv) **Foliar application or spraying of nutrient solutions:** Some soluble fertilizers viz., urea, micro-nutrients and others are dissolved in water, filtered and sprayed over the crop foliage with the help of a suitable sprayer because of their high solubility or low requirement (micro nutrients). In this method, crop plants respond to nutrients quickly (within 24 hours of application) to overcome any shock. The soil reaction, topography and texture do not cause adverse effect on the nutrient availability. However, uneven spraying and improper concentration of the solution may lead to hazardous effect over the crop. In case of urea, only fresh urea should be used to avoid toxic effect of biuret. Maximum permissible limit of biuret concentration in urea is 1.5 percent while in foliar spray it must not exceed 0.25%. DAP can also be used for foliar spray. Low temperature and high humidity are required for nutrient absorption. So morning and evening time are most suitable for their application. Right stage of foliar spraying is at flowering stage. For most nutrients, the spray concentration should not exceed 3 to 5% when conventional high volume sprays are used. Concentration varies with chemical nature and crop tolerance. The rate of absorption and effectiveness depends on solubility of compounds, ionization, activating molecules and presence of other nutrients.

Fertilizers commonly used for foliar application

N : Urea, Calcium nitrate

P : Super phosphate, Potasium monophosphate, orthophosphoric acid

K : KCl, KNO_3

Ca : $CaCl_2$

Mg : $MgSO_4$

Fe : Ferrous sulphate, Fe chelate

Mn : $MnSO_4$

Zn : $ZnSO_4$, Zn chelates

Mo: Ammonium molybdate

B : Boric acid, Borax

METHOD OF BIOFERTILIZER APPLICATION

Biofertilizers by rendering unavailable sources of elemental nitrogen, bound phosphates and decomposed plant residues into available forms helps in enhancing soil fertility and increasing crops yields. The bio-inoculants or preparations containing microorganisms that supply nutrients especially nitrogen and phosphorus are known as biofertilizers. On the basis of nutrients supply, these are broadly classified in four groups: (i) nitrogen fixers (ii) phosphorus solubilizers (iii) plant growth promoters and (iv) organic matter decomposers as discussed in chapter 7.

Methods for treating the seeds with inoculations are as follows

(i) **Seed treatment for *Rhizobium* and phosphorus solubilizing bacteria:** Roughly 500g *Rhizobium or PSB* culture is required for treating seeds for an area of 1 ha. Thick slurry of carrier based inoculums is prepared by mixing 500 g of inoculums in 1.25 L of water. Stickiness can be improved by using 10% jaggery solution or arabic gum. The required quantity of seed is mixed with the slurry and dried in shade and sown within 24 h. Under no circumstances seeds should be dried in direct sunlight because seed coat will be destroyed by UV rays and it will destroy the inoculated bacteria. Sometimes the rate of application is 100 gm biofertilizer for 10 Kg of seed.

(ii) **Seedling treatment:** Mix 500 g culture in 2.5 L water. Dip the roots of the seedling in the solution for 20 minute prior to transplanting.

(iii) Soil application: Mix biofertilizer along with organic manure in the ratio 1:25 and apply in the field.

Precautions

- *Rhizobium* strains are specific and different for different crops (Annexure 10), therefore, select appropriate strain for better results.
- It should contain at least minimum quantity of microorganism per g of inoculants.
- Microorganism should be viable.
- The inoculation should be kept in cool and dry place.
- It should not be expired.

■■■■

9

NUMERICAL EXERCISES ON FERTILIZER/ MANURE REQUIREMENT

INTRODUCTION

The calculation for requirement of fertilizer sources to supply the recommended nutrient requirement of crops has a great impact on balanced availability of nutrients for higher recovery besides reducing the cost of production. As pointed out in previous exercise regarding the importance of four R's of fertilizers and manures, the calculation of fertilizer requirement is of prime importance for higher productivity without affecting the soil and environmental health.

The fertilizer requirement changes with soil types, previous crop, season, cropping system, cropping situation, purpose of growing, source of fertilizer etc. The knowledge regarding nutrient content in different sources is essential for calculating the fertilizer requirement and discussed in chapter 7.

Table 1. Conversion of nutrients in their different forms

Organic matter	=	Organic carbon x 1.724 (Organic matter contains 58 % organic carbon)
Protein content (%)	=	% N x 6.24
NO_3%	=	N% x 4.54
N%	=	NO_3% x 0.22
P_2O_5%	=	P% x 2.29
P%	=	P_2O_5% x 0.44
K_2O%	=	K% x 1.20
K%	=	K_2O% x 0.83
CaO%	=	Ca % x 0.40
Ca%	=	CaO x 0.71
MgO%	=	Mg % x 1.63
Mg%	=	MgO % x 0.61

Principles involved in calculating the fertilizer requirement are as follows:

1. First of all calculate the quantity of nutrients supplied by bulky and concentrate manures.
2. Calculate the quantity of major nutrients (primary nutrients) supplied by straight and complex fertilizer (e.g. Calculate N from urea, P_2O_5 from DAP and K_2O from MOP followed by second primary nutrient i.e. N in case of DAP)
3. Deduct the quantity of second primary nutrient supplied by complex fertilizer from the recommended quantity of fertilizers to fulfill the requirements of particular nutrient such as N supplied by DAP followed by remaining N supplied by urea or other straight fertilizer (Example 2).

Formula for calculating the quantity of nutrient sources (fertilizer or manure) required to supply recommended dose of nutrient (kg)

$$\text{Quantity of fertilize manure required (kg)} = \frac{\text{Quantity of nutrient recommended (kg)}}{\text{Percent content of nutrient in source (\%)}} \times 100$$

Formula for calculating the quantity of nutrient supplied by nutrient sources (kg) (In case of complex fertilizer)

$$= \frac{\text{Nutrient content in applied source (\%)} \times \text{Quantity of nutrient required (dose of recommendation) (kg)}}{100}$$

Example

A farmer wants to grow gram crop in 4 hectare area. The sources of nutrients are Urea, SSP and MOP to supply N, P_2O_5 and K_2O, respectively. The recommended dose for N, P_2O_5 and K_2O is 30.0, 60.0, 20.0 kg/ha respectively. All the fertilizers will be applied at the time of sowing.

Hint: Here, Urea, SSP and MOP contain 46, 16 and 60% of N, P_2O_5 and K_2O, respectively.

Solution

1. Quantity required for 4 ha

N= 30 x 4 = 120 kg.

P_2O_5= 60 x 4= 240 kg.

K_2O= 20 x 4= 80 kg.

Sol. For Nitrogen

1st Method

$\because$ 46 kg N will supplied from urea = 100 kg

$\therefore$ 1 kg N is received from urea = 100/46 kg

$\therefore$ 120 kg N is received from urea = 100/46 x 120=260.87 = 261kg

2nd Method

$$\text{Quantity of urea required to supply 120 kg N} = \frac{\text{Quantity of nutrient required (kg)}}{\text{Percent of nutrient conten in source (\%)}} \times 100$$

$$\text{Quantity of urea required (kg)} = \frac{120}{46} \times 100 = 260.87 = 261$$

For P_2O_5

$\because$ 16 kg P_2O_5 is supplied from SSP = 100 kg

$\therefore$ 1 kg P_2O_5 is supplied from SSP = 100/16 kg

$\therefore$ 240 kg P_2O_5 is supplied from SSP = 100/16 × 240=1500.0 kg

$$\text{Quantity of SSP required (kg)} = \frac{240}{16} \times 100 = 1500\ \text{kg}$$

For K_2O

$\because$ 60 kg K_2O is received from MOP = 100 kg

$\therefore$ 1 kg K_2O is received from MOP = 100/60 kg

$\therefore$ 80 kg K_2O is received from MOP = 100/60 × 80=133.3= 133 kg

$$\text{Quantity of MOP required (kg)} = \frac{80}{60} \times 100 = 133.3\ \text{kg} = 133\ \text{kg}$$

Interpretation

For growing 4 ha gram crop, he needs to purchase 261, 1500 and 133 kg of urea, SSP and MOP, respectively.

Example 2

A farmer wants to cultivate potato crop, the recommended dose is 180: 80: 120 kg N, P_2O_5 and K_2O respectively. The sources of nutrient are urea, DAP and SOP, respectively. Calculate the quantities of each fertilizer required to fulfil the requirement of nutrients for one ha.

Hint: Urea contain 46% N, DAP contain 18% N and 46% of P_2O_5 and SOP contain 48% K_2O.

Solution

(First of all calculate the quantity of phosphorus supplied by DAP)

$$\text{Quantity of DAP required (kg)} = \frac{80}{46} \times 100 = 173.9 = 180\ \text{kg}$$

Calculating the quantity of Nitrogen supplied by 180 kg DAP

$$= \frac{\text{Percent of nutrient content in DAP} \times \text{Quantity of DAP}}{100}$$

$$= \frac{18 \times 180}{100} = 32.4 \text{ kg}$$

180 kg DAP supplies 32.4 kg N, therefore, the remaining quantity of N i.e. 180.0-32.4 = 147.6 kg will have to be supplied by urea thus

$$\text{Quantity of urea required (kg)} = \frac{147.6}{46} \times 100 = 320.87 = 321 \text{ kg}$$

$$\text{Quantity of SOP required (kg)} = \frac{120}{50} \times 100 = 240 \text{ kg}$$

Interpretation

For growing potato in 1 ha, farmer will need to purchase 321, 180 and 240 kg of urea, DAP and SOP, respectively.

Problem 3

Calculate the total quantity of urea, SSP and MOP for 1.0 ha for barley crop. The application ratio is 80:40:40. Farmer applied 5 t ha of FYM per ha at the time of last ploughing.

Hint:

(i) FYM contains 0.5, 0.25 and 0.5% N, P_2O_5 and K_2O, respectively.

(ii) Urea, SSP and MOP contain 46, 16 and 60% of N, P_2O_5 and K_2O, respectively

Solution:

Calculating the quantity of nutrient supplied by 5000 kg FYM

$$\text{Quantity of N supplied (kg)} = \frac{\text{Percent of N content in FYM} \times \text{Quantity of FYM Supplied (kg)}}{100}$$

$$= \frac{0.5 \times 5000}{100} = 25.0 \text{ kg}$$

$$\text{Quantity of } P_2O_5 \text{ supplied (kg)} = \frac{\text{Percent of } P_2O_5 \text{ content in FYM} \times \text{Quantity of FYM Supplied (kg)}}{100}$$

$$=\frac{0.25\times5000}{100}=12.5\text{ kg}$$

$$\text{Quantity of }K_2O\text{ supplied (kg)}=\frac{\text{Percent of }K_2O\text{ content in FYM}\times\text{Quantity of FYM Supplied (kg)}}{100}$$

$$=\frac{0.5\times5000}{100}=25.0\text{ kg}$$

The 5 t of FYM supplies 25.0, 12.5 and 25.0 kg of N, P_2O_5 and K_2O, respectively, therefore, the remaining quantity of N i.e 80.0-25.0 = 55.0 kg, P_2O_5 i.e. 40.0-12.5 = 27.5 kg and K_2O i.e. 40.0-25.0 = 15.0 kg will be supplied by urea, SSP and MOP.

$$\text{Quantity of urea required (kg)}=\frac{55.0}{46}\times100=119.56=120\text{ kg}$$

$$\text{Quantity of SSP required (kg)}=\frac{27.5}{16}\times100=171.9=172\text{ kg}$$

$$\text{Quantity of MOP required (kg)}=\frac{15}{60}\times100=25\text{ kg}$$

Interpretation

For growing barley in 1 ha, farmer needs to purchase 120, 172 and 25 kg of urea, SSP and MOP, respectively.

Example 4

Calculate the total N, P_2O_5 and K_2O for wheat crop, if it to be was fertilized with 250 kg urea, 480 kg SSP and 60 kg MOP

Hint: Quantity of fertilizers given i.e 250 kg urea, 480 kg SSP and 60 kg MOP. Urea contains 46% N, SSP contains 16% P_2O_5 and MOP contains 60% K_2O.

Solution:

Quantity of N supplied by urea

100 kg urea supplies N = 46 kg

1 kg urea supplies N = 46/100 kg

250 kg urea supplies N = 46/100 × 250=115 kg N

Or

The quantity of nitrogen supplied by 250 kg urea

$$= \frac{\text{Percent of nutrient content in urea} \times \text{Quantity of urea}}{100}$$

$$= \frac{46 \times 250}{100} = 115 \text{ kg}$$

Quantity of P_2O_5 supplied by SSP

$\because$ 100 kg SSP supplies P_2O_5 = 16 kg

$\therefore$ 1 kg SSP supplies P_2O_5 = 16/100 kg

$\therefore$ 480 kg SSP supplies P_2O_5 = 16/100 × 480=76.8 kg : 77 Kg

Or

The quantity of phosphorus supplied by 480 kg SSP

$$= \frac{\text{Percent of nutrient content in SSP} \times \text{Quantity of SSP}}{100}$$

$$= \frac{16 \times 480}{100} = 76.8 = 77.0 \text{ kg}$$

for K_2O

$\because$ 100 kg MOP supplies K_2O = 60 kg

$\therefore$ 1 kg MOP supplies K_2O = 60/100 kg

$\therefore$ 60 kg MOP supplies K_2O = 60/100 × 60= 36 kg

Or

The quantity of potash supplied by 40 kg MOP

$$= \frac{\text{Per cent of nutrient content in MOP} \times \text{Quantity of MOP}}{100}$$

$$= \frac{60 \times 60}{100} = 36.0 \text{ kg}$$

Interpretation

The wheat crop sown by farmers has been nourished with 115, 77 and 36 kg of N, P_2O_5 and K_2O respectively.

■■■■

10

IDENTIFICATION OF PESTICIDES

Introduction

Different pests like insects, weeds and microbes (Fungi, virus, bacteria etc.) cause a considerable loss to crop economic yields and deteriorate the quality of produce. Pesticide is a chemical substance used for destroying different pests causing damage to crop. These pesticides can be classified by target organisms and have been depicted in Table 1. The first major commercial pesticide were an arsenical compound, Paris green (1867), Bordeaux mixture (1885), lime sulphur (1880), lead arsenate (1892) and nicotine sulphate (1909). However, a rapid increase in their use began in the 1930s and 1940s following the discovery of synthetic organic chemicals that could be used to control plant diseases caused by fungi (e.g., thiram in 1931), insects (e.g., DDT in 1939), or weeds (e.g., 2,4-D in 1942). In India, 248 pesticides have been registered for use against different pests, diseases and weeds, among which 88 are insecticides alone. Pesticides are seldom used in pure form since only small quantities of toxicant are required to kill the pests. Hence, the toxicant i.e. active ingredient is mixed with inert materials known as additives or adjuvant to facilitate spraying lenatheu and to increase their toxic effects. The mixture thus formed is called a formulation. Pesticides formulations are of three types: solid, liquid and gaseous. Thus, it is most urgent to learn about pesticides formulation, application methods, and disposal techniques for higher control efficiency and to protect the grower against any causality. The effectiveness to these pesticides varies and is mainly influenced by prevailing climate conditions. Always keep in mind, protect the beneficial insects, parasites and their parasitoids and reduce the population of pests.

Pesticides are always applied on the basis of soil and foliage on the site, time of control, nature of pests and stage of damage. The method of application viz., spraying, broadcasting, injecting, spreading etc. depends on types of formula-tions. The higher efficiency may be achieved by calculating right dose based on stage of crop, area to be treated and selection of nozzles. The ideal pesticides must control the population of individual pests and side by side repelling parasitoids, interference and drift or runoff may affect untreated pests and plants on nearby plots.

Types of pesticides formulations

I. Solid Formulations

1. **Dust (D):** Lindane 3% dust.
2. **Granule (G):** Furadan 3 G, Ekalux 5G. Denthiocarb
3. **Wettable powder (WP):** Sevin 50 WP. Atrazine
4. **Water dispersible powder (WDP):**

II. Liquid Formulations

1. **Emulsifiable concentrate (EC):** Ekalux 25 EC. Malathion 50 EC. 2,4-D.
2. **Solution:**
3. **Water soluble concentrate (WSC/SL):** Dimecron and Nuvacron
4. **Aerosols:**

III. Gaseous formulations

1. **Fumigants:** Celphos (tablet form), Methyl bromide (liquid form).

Statutory information furnished along with pesticides

Every pesticide container should have a label with following information which indicates the nature of pesticide and other useful information. The following information must be furnished on the label.

- Name of pesticide (brand name, trade name and common name)
- Name of the manufacturer and address
- Registration number
- Kind and name of active ingredient
- Type of formulation (EC, WP, granule etc.)
- Net contents by weight
- Batch number assigned by the manufacturer
- Date of manufacture
- Expiry date
- Antidote statement

- Warning symbol and signal (warning symbol is of diamond shape consisting of two triangles with a color in the lower triangle and a signal in the upper triangle)
- Name of the pests against which the chemical may be used.
- Directions for use.
- Warning and cautioning statements, symptoms of poisoning, antidotes and first aid.
- Directions for storage, careful handling and method of disposal.

Classification based on toxicity: The pesticides have been classified into four categories based on extent of toxicity and presented in tabular form with related information (Table 1).

Table 1. Pesticides classification based on toxicity

Toxicity class	**Extremely toxic**	**Highly toxic**	**Moderately toxic**	**Slightly toxic**
Depiction		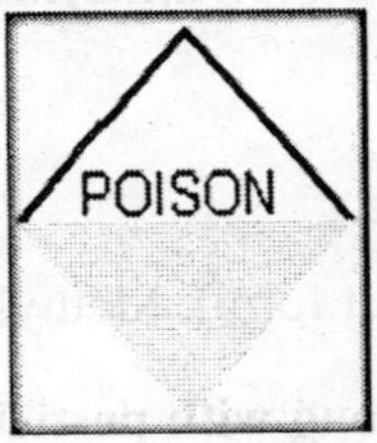		
Colour of lower triangle	Bright red	Bright yellow	Bright blue	Bright green
LD 50 value (mg/kg body weight)				
Oral	1-50 mg/kg	51-500 mg/kg	501-5000 mg/kg	> 5,000 mg/kg
Dermal	1-200 mg/kg	201-2,000 mg/kg	2001-20,000 mg/kg	> 20,000 mg/kg
Signal words upper half	POISON (In red)	POISON (In red)	DANGER	CAUTION
Warning words (Outside the diamond)	Keep out of reach of children, if swallowed (or) if symptoms of poisoning occur, call physician immediately	Keep out of reach of children	Keep out of reach of children	Safe pesticide, harmless, non-poisonous

Table 2. List of Common Conventional Insecticides

Generic name	Common formulations	Recommended dose Rice (R)/ Other crops (OC)	Remarks
Cyclodienes			
Endosulfan (Banned now)	Thiodan 35 EC/AF Endocel 35 EC/AF	0.05%	Broad spectrum, relatively safer to beneficial insects
	Hexasulfan 4 DP	2 kg ai/ha (OC)	For pests of vegetables
Carbamates			
Carbaryl	Sevin 5 DP Sevin 10 DP	2 kg ai/ha (R, OC)	Broad spectrum insecticidal action.
	Sevin 50 WP	0.15-0.2%	
Carbofuran	Furadan 3 G	0.5 kg to 0.75 kg ai/ha	For pest control in rice, soil borne pest
Organophosphates			
Fenitrothion	Sumithion 50 EC/AF	0.05%	Contact and stomach action. Broad spectrum
Mercaptothion	Malathion 50 EC/AF	0.10%	Safe for controlling pests of vegetables and storage pests
DDVP	Vapona 65 EC/AF Nuvan 100 EC/AF	0.05%	Contact and fumigant action. Least residual toxicity. Safe to be applied on vegetables
Quinalphos	Ekalux 25 EC/AF	0.025 -0.05%	Broad spectrum toxicity
Phosalone	Zolone 35 EC/AF	0.07%	Broad spectrum insecticide cum acaricide
Fenthion	Lebaycid 50 EC/AF	0.05%	For rice stem borer and other pests of rice
Dimethoate	Rogor 30 EC	0.03 – 0.05%	Systemic insecticide cum acaricide
Monocrotophos	Nuvacron 40 EC/AF Monocil 40 EC/AF	0.05%	Systemic, persistent. Long residual action.
Phosphamidon	Dimecron 86 EC/AF	0.05%	Systemic with weak contact toxicity. Insecticide cum miticide
Chlorpyriphos	Dursban 20 EC/AF	0.02%	Effective against a range of pests and termites

Generic name	Common formulations	Recommended dose Rice (R)/ Other crops (OC)	Remarks
Triazophos	Hostathion 40 EC/AF	0.04%	Effective against rice leaf folder
Acephate	Starthene 75 SP	0.05%	Effective against aphids, leaf miners, caterpillars

Newer Molecules

Generic Name	Type of formulations	Trade name
Carbamates		
Carbosulfan	25 DS, 25 EC, 6 G	Marshal
Thiocarb	75 WP	Larvin
Indoxacarb	14.5 SC	Avaunt
Organophosphates		
Fenazaquin	50 EC	Magister
Profenofos	50 EC	Curacron
Synthetic Pyrethroids		
Beta Cyfluthrin	2.5 E	Bulldock
Lambda cyhalothrin	5 EC 2.5 EC	Karate Kungfoo
Alphamethrin	10 EC	Gem
Neonicotinoids		
Imidacloprid	17.8 EC, 70WS	Confidor, Gaucho
Acetamiprid	20 SP	Pride
Thiamethoxam	25 WG 70 WS	Actara Cruiser
Organic insecticides		
Abamectin	1.8 EC	Vertimec
Spinosad	45 SC 2.5 SC	Tracer Success
Cartap	50 SP	Caldan
Insect growth regulators		
Lufenuron	5 EC	Match

Generic Name	Type of formulations	Trade name
Teflubenzuron	15 SC	Nomolt
Novaluron	10 EC	Rimon
Difenthiuron	50 SC	Polo
Biopesticides		
Azadirachtin	0.15% WW Min.	
Bacillus thuringiensis	5% WP	
Insecticide Mixtures		
Profenophos 40%+ Cypermethrin 4%	EC	Polytrin 44
Cypermethrin 5% + Chlorpyriphos 50%	EC	Nurelle D 505
Cypermethrin 5% + Quinalphos 20%	EC	Prachand
Deltamethrin I% + Triazophos 35%	EC	Spark
Fenvalerate 3% + Acephate 25%	EC	Koranda

Table 3. Fungicides, their formulations and trade names

FUNGICIDES			
	Common name	**Chemical composition/ Doses**	**Trade Name**
COPPER FUNGICIDES			
Copper Sulphate compound	Bordeaux mixture	It is prepared by suspending 5 Kg of copper sulphate and 5 Kg of lime in 500 liters of water (1%)	–
	Bordeaux paste	It is prepared by mixing 1 Kg of copper sulphate and 1 Kg of lime in 10 liters of water	–
	Burgundy mixture	Sodium carbonate is used in place of lime. It is prepared by mixing 1 Kg of copper sulphate and 1 Kg of sodium carbonate in 100 liters of water	–
	Cheshunt compound	It is a compound prepared by mixing 2 parts of copper sulphate and 11 parts of ammonium carbonate	–
	Chaubattia paste	It is a compound prepared by mixing 800g of copper sulphate and 800g of red lead in 1 liter of lanolin or linseed oil	–
Other	Copper oxy chloride	0.3 to 0.5% for foliar application, 25 to 35 Kg/ha for dusting	Blitox-50, Bluecopper- 50, Cupramar-50

FUNGICIDES

	Common name	Chemical composition/ Doses	Trade Name
COPPER FUNGICIDES			
	Cuprous oxide	0.3% for foliar spray	Fungimar & Perenox
	Copper hydroxide	0.3% for foliar spray	Kocide
SULPHUR FUNGICIDES			
INORGANIC			
	Lime sulphur	10-15 liters in 500 liters of water	
	Sulphur dust	4-5g/Kg seed for ST, 10-30 Kg/ha for dusting on crops, 100 Kg/ha for soil application	Kolodust, Mico- 999
	Wettable sulphur	0.2-0.4 % for foliar spray	Sulfex, Thiovit, Cosan
ORGANIC	**Di-alkyl Dithiocarbamates**		
	Ziram	0.15 to 0.25% for foliar spray	Ziride, Hexazir, Milbam, Zerlate
	Ferbam	0.15 to 0.25% for foliar spray	Coromet, Ferbam, Fermate, Fermocide, Hexaferb, Karbam, Black
	Thiram	0.15 to 0.2% as foliar spray, 0.2-0.3% as dry seed treatment, 15-25Kg/ha as soil application	Arasan, Hexathir, Tersan, Thiram, Thiride
	Monoalkyl dithyiocarbamates		
	Nabam	0.2% as foliar spray	Chembam, Dithane D-14, Dithane A-40 and Parzate liquid
	Zineb	0.1 to 0.3% for foliar application	Dithane Z-78, Hexathana, Lanocol and Parzate
	Vapam or Metham sodium	1.5 to 2.5 liters per 10 m^2 area	Chem-vape, vapam, vitafume, VPM
	Maneb	0.2% to 0.3% as foliar application	Dithane M 22, Manzate and MEB. Mancozeb (78% Maneb + 2% zinc ion) Dithane M 45, Indofil M 45
HETROCYCLIC NITROGENOUS COMPOUNDS			
	Captan (Kittleson's killer)	0.2 to 0.3% for dry seed treatment, 0.2 to 0.3% for foliar	Captan 50W, Captan 75 W, Esso

FUNGICIDES			
	Common name	**Chemical composition/ Doses**	**Trade Name**
COPPER FUNGICIDES			
		spray, 25 to 30 Kg/ha for furrow application	fungicide, Orthocide 406, Hexacap, Vancide 89
	Folpet	0.1 to 0.2% for spraying	Phaltan
	Captafol	0.15 to 0.2% for spraying, 0.25% for seed treatment, 0.15% for soil drenching	Difosan, Difolaton, Sanspor, Foltaf
	Iprodione	0.1 to 0.2% for foliar application	Rovral, Glycophene
	Vinclozolin	0.1 to 0.2% for foliar application	Ornalin, Ronilan, Vorlan
MISCELLANEOUS FUNGICIDES			
	Chlorothalonil	0.2 to 0.3% for foliar application	Bravo, Daconil, Kavach, Thermil, Exotherm
	Dinocap	0.1 to 0.2% for spraying	Karathane, Arathane, Capryl, Mildex, Mildont and crotothane
	Dodine	0.075% for spraying	Cyprex, Melprex, Guanidol and Syllit
SYSTEMIC FUNGICIDES			
ACYLALANINES	Metalaxyl	3-6 g/Kg seed for seed treatment, 1 to 1.5 Kg a.i/ha for soil application, 0.1 to 0.2% for foliar spray	Ridomil 25 % WP, Apron 35 SD, Ridomil MZ-72WP
	Benalaxyl	0.1 to 0.2% for foliar spray, 1 to 1.5 Kg a.i/ha for soil application	Galben 25% WP and 5% G
AROMATIC HYDROCARBONS	Chloroneb	0.2% for seed treatment	Demosan
BENZIMIDAZOLES	Carbendazim	0.1% for foliar spray, 0.1% for soil drench, .25% for ST, 500- 000 ppm for post-harvest dip of fruits	Bavistin50WP, MBC, Derosol 60WP,Agrozim, Zoom
	Benomyl	0.1 to 0.2% for ST, 50-60g/100 L for foliar spray, 50-200ppm for soil drenching, 12-45 Kg a.i/ha for soil broadcast, 100- 500 ppm for post harvest fruit dip	Benlate 50WP
	Thiabendazole	0.2 to 0.3% for spraying, 1000 ppm for fruit dip	Mertect 60WP, Mycozol, Arbotect, Tecto and Storite
ALIPHATICS	Prothiocarb	5.6 Kg a.i/ha for soil application	Previcur, Dynone
	Propamocarb	3.4 and 4.8 Kg a.i/ha for soil	Previcur-N,

FUNGICIDES

	Common name	Chemical composition/ Doses	Trade Name
COPPER FUNGICIDES			
		application	Dynone-N, Prevex, Benol
OXATHINS or CARBOXI-MIDES	Carboxin	0.15 to 0.2% for seed treatment, 0.5% for spraying	Vitavax 75WP, Vitaflow
	Oxycarboxin	0.1 to 0.2% for foliar spray, 0.2 to 0.5% for ST	Plantavax 75 WP, Plantavax 20EC, Plantavax 5% liquid
IMIDAZOLES	Imazalil	0.1% as post harvest dip	Fungaflor, Bromazil and Nuzone
	Fanapanil	0.05% foliar spray	Sistane 25 EC
MORPHO-LINES	Tridemorph	0.1% for foliar spray	Calixin 75EC,Bardew, Beacon
ORGANO-PHOSPHATES	Iprobenphos	30-45 Kg of granules/ha, 1 to 1.5 liters of 48% EC in 1000 ml of water for foliar spray	Kitazin 48EC, Kitazin17G, Kitazin 2% D
	Ediphenphos	400 to 500 ppm for spraying, 30 to 40 Kg/ha for soil application	Hinosan 30 and 50% EC, Hinosan 2%D
ALKYL PHOSP-HONATES	Fosetyl-Al or Aluminium Tris	0.15% for foliar spray, 0.2% for soil drench	Aliette 80WP
PYRIMIDINES	Fenarimol	2g/Kg seed as ST, 20 to 40 ml/ 100 liters of water for spraying	Rubigan 50% WP, 12%EC
THIOPHA-NATES	Thiophanate	0.1 to 0.2% for spraying	Topsin50WP, Cercobin 50WP
	Thiophanate methyl	% for spraying	TopsinM70WP, Cercobin M 70WP
TRIAZOLES	Triadimefon	0.1 to 0.2% for spraying, 0.1% for seed treatment	Bayleton, Amiral
	Tricyclazole	2g/Kg for seed ST, 0.06% for spraying	Beam 75WP, Baan 75WP, Trooper 75WP
	Bitertanol	0.05 to 0.1% for foliar spray	Baycor and Sibutol
	Hexaconazole	0.2% for spraying	Contaf Anvil 5% EC

	Common name	Chemical composition/ Doses	Trade Name
FUNGICIDES			
COPPER FUNGICIDES			
	Propiconazole	0.1% for foliar application	Tilt, Desmel 25% EC
	Myclobutanil	0.1 to 0.2% for spraying	Systhane 10WP
STROBILURINS	Azoxystrobin	0.1 % for spraying	Amistar, Quadris
	Kresoxim methyl	0.1% for spraying	Ergon, Stroby Discus

Table 4. List of antibiotics

	Antibacterial antibiotics	
1. Streptomycin sulphate	100-500 ppm	Agrimycin,-100, Streptomycin sulphate, Plantomycin, Streptocycline, Paushamycin, Phytostrip, Agristrep and Embamycin
2. Tetracyclines	All these antibiotics are bacteriostatic, bactericidal and mycoplasmastatic.	Terramycin or Oxymicin (Oxytetracycline)
	Antifungal antibiotics	
1. Aureofungin	Normally sprays at 50-100 ppm.	Aurefungin-Sol. Containing 33.3% Aureofungin
2. Griseofulvin		Griseofulvin, Fulvicin and Grisovin
3. Cycloheximide		Actidione, Actidione PM, Actidione RZ and Actispray
4. Blasticdin		Bla-s
5. Antimycin		Antimycin
6. Kasugamycin		Kasumin
7. Thiolution		–
8. Bulbiformin		–
9. Endomycin		–
10. Nystatin		Mycostain and Fungicidin
11. Eurocidin		Pentaene G-8

Table 5. Nematicides (Fumigants and Non-fumigants)

Fumigants

D-D (Mixture of 1,2-dichloropropane and 1,3-dichloropropene)

1,3-D (1,3-Dichloropropene)

EDB- (Ethylene Dibromide)

DBCP (1,2-Dibromo-3-Chloropropane)

Methyl Bromide

Chloropicrin

MITC (Methyl isothiocyanate)

Metam Sodium

Dazomet

Sodium Tetrathiocarbonate

(Non-fumigants) Carbamates

Aldoxycarb

Carbofuran

Oxamyl

(Non-fumigants) Organophosphates

Ethoprop

Fenamiphos

Cadusafos

Fosthiazate

Terbufos

Fensulfothion

Thionazin

■■■■

11

IDENTIFICATION AND WORKING OF DIFFERENT TILLAGE IMPLEMENTS

Introduction

Soil is the medium in which crops are grown but in its natural state, it is not in an ideal condition to grow them satisfactorily for tender shoots of germinating seeds to push above the soil surface without any difficulty and the young roots penetrate easily into the lower layers of soil in search of food, water and air. So soil should be soft and friable, free from weeds and have sufficient water and air for optimum germination emergence and overall crop growth. Such ideal condition of soil can be achieved by manipulating the soil properly and bringing it in good tilth through a series of mechanical operations by tillage implements. These operations may be very easy or may be hard. Tillage operations mostly followed are ploughing, clod crushing, dicing, harrowing, levelling, compacting, interculturing etc. Soil tilth is the term used to express soil condition resulting from tillage. The identification of different farm implements commonly used for tillage operation and knowledge about their working principles is very important for smooth and timely completion of field operations efficiently.

Tillage implements can be broadly classified based on power used for working:

- Manually operated or hand tools: In agriculture certain operations are done manually with the help of small implements e.g. fork, hand hoe, sickle, spade
- Animal drawn (eg., desi plough, bund former, interculture implements, wooden planks etc.).
- Mechanically operated, i.e, drawn by machine – tractor drawn. e.g. disc plough, disc harrow, cultivator, rotovator.

Based on the type of tillage for which they are used, the implements may be:

- Primary tillage implements: The implements which are used for ploughing, are known as primary tillage implements. Disc plough, mould bold plough etc are the examples of primary tillage implements.

- Disc harrows, cultivators, blade harrow etc. are the examples of secondary primary tillage implements
- Intercultural implements.

(a) Primary tillage: It mainly includes the ploughing operation which is opening of the compacted soil with the help of different ploughs. Ploughing is done to:

(1) Open the hard soil,

(2) Separate the top soil from lower layers,

(3) Invert the soil whenever necessary and

(4) Uproot the weeds and stubbles.

The cutting and inverting of the soil that is done after the harvest of the crop or untitled fallow or to bring virgin or new land under cultivation is called primary tillage. It may be done once or twice a tear in normal or settled agriculture or once in four to five years in dry land agriculture.

(b) Secondary tillage : Lighter or finer operation performed on the soil after primary tillage are known as secondary tillage which includes the operations performed after ploughing, leveling, discing, harrowing etc.

(c) Inter tillage/ Inter cultivation/ Interculture/ after care operation: The tillage operations that are carried out in the standing crop are called inter tillage operations. The tillage operation done in the field after sowing or planting and prior to the harvesting of crop plants known as inter cultivation. It includes gap filling, thinning, weeding, mulching, top dressing of fertilizers, hoeing, earthling up etc. unless these are carried out at right time, with suitable implements mainly hoes and hand tools the crop will not attain a vigorous growth. These operations are carried out in between the crop rows.

The common implements used on all types of soils are described below:

Country/wooden/Desi plough: A country plough has the following parts- (1) share (2) shoe (3) body (4) handle and (5) beam. The country plough has certain modifications from place to place in India. A small piece of flat iron serves as the share, the piercing point of the plough. The share is fixed on the shoe. The beam attaches the body of the plough and the yoke. The beam end is provided with grooves for proper hitching. The handle is fixed behind the body to direct the movement of the plough. The ploughs are worked usually with a pair of bullocks. When the plough is worked in the soil, the shoe and share enter the soil under the wedge action of the body and makes 'V' shaped furrows. The furrows usually would have a depth of 10 cm and a width of 15 cm on the top of the furrows. It covers 0.15 to 0.20 ha in 8 h.

The mould board plough: A mould board plough, in general, is having a plough bottom, a standard, a handle and a beam. The share is that part of the plough, which actually cuts the soil in a horizontal direction and separates the furrow slice from the ground. This separated furrow slice is transferred through the side of the landside to the mould board. After the completion of ploughing, there will be a furrow in double the size of a plough furrow in the centre. This is known as *dead furrow.*

Disc plough: In a disc plough, instead of mould board and share, the plough bottom consists of one or more steel concave discs with their plane of cutting edge set at an angle with the direction of travel. The discs are mounted on a frame. The discs cut the soil, break it, and push it side ways. Penetration of soil is effected mainly due to the weight of the plough and the angle of the disc.

Cultivator: Cultivators perform functions intermediate between those of ploughs and harrows. A cultivator is operated by the action of a number of tynes, which break up the soil, but perform no inversion. Cultivators are usually classified into three by the design of their tynes; the *rigid tyne*, the *spring-loaded tyne* and the *spring tyne* cultivator.

Rotary cultivator: Rotary cultivators are now widely used as an alternative to ploughing. They are capable of producing a seedbed in a fewer operations than ploughing. They produce the combined effect of ploughing and cultivating. Rotary cultivators can be either *pedestrian-operated* or *tractor mounted.* The implement essentially consists of a rotating shaft- the rotor- on which are bolted a series of 14-20 'L' shaped blades, which can be changed when worn. The shaft and the tynes together are called the *rotovator.* There is a shield around the rotor which controls the type of tilth produced, and protects the operator from flying stones and clods. The blades rotate in the direction of forward travel and cut out slices, which are pulverised against the shield and by other blades. The depth of cut is controlled by an adjustable skid, which enters the soil and prevents the rotovator going too deep.

Spike tooth harrow: These types of harrows are mainly used for seedbed preparation of rice. On a wooden frame, about 30-35 wooden spikes are fitted. The beam is fixed at the centre of the frame. The harrow is drawn by yoking to a pair of bullocks. The depth of penetration varies from 5 to 7.5 cm.

Blade harrow (Bakhar, Guntaka): They are popular in North Indian States. Blade harrows are generally used in clayey soils. It consists of a blade like cutting edge, which is fixed horizontally. It brings about pulverisation of about 5-8 cm of topsoil. Certain types are used for the harvesting of groundnut also.

Disc harrow: Disc harrows can be conveniently used as an alternative to cultivators, when the surface trash is more. They consist of a series of concave saucer shaped discs mounted on a shaft. All the discs mounted on a shaft are together called a *gang.* Two or four gangs will be mounted on a frame as a set. In certain disc harrows, the discs may be having notches. These types of notched disc harrows are more suitable in grassy situations. Based on the number of gangs and their settings on the frame, disc harrows can be *single action, double action* or *offset.* In single action types, two gangs are placed end to end. Both gangs throw soil either towards the centre or away from the centre, based on the fixing of discs on the frame. In the double action types, there are four gangs. The discs on the two gangs placed in the front would be in one direction and the discs on the two gangs placed behind in the opposite direction. In the offset type, there are only two gangs placed one behind the other.

Digging hoe: It is the most familiar and widely used multipurpose tillage implement used almost all over country for intercultivation (hoeing and weeding) in crops. It is also used for digging, making bunds, furrows, ridges, etc. and for harvesting in crops like sweet potato, potato, elephant foot, yam, ginger and turmeric. It consists of a wooden handle of varying length to which a flat iron or steel blade is attached at nearly right angles. The operator lifts the implement with both hands and by the application of force the cutting edge penetrates into the soil.

Spade: It is prepared from iron sheet and it is about square shaped. Its one edge is sharp, while its opposite side is blunt with hole in the middle for fitting wooden handle.The spade is mainly used for removing soil, for preparing seedbed, canals and bunds, for digging and trenching in gardens and for giving irrigation and other such agricultural operation. Spade handles are made out of light, but strong wood generally from *babul*.

Pick-axe *(Koonthali*): Pick–axe is an implement very commonly used in road-making especially to dig hard pans and stony lands. One end of the pick-axe is pointed and the other slightly flat. In agriculture, it is used for breaking the hard and compact soil especially laterite pans, particularly when digging pits for tree crops.

Fork: A fork is like the human hand in shape, with three to five strong pointed prongs or tynes made up of wood or iron. The prongs may be 10 cm to 20 cm long or sometimes more and are fixed to a handle. It is used for collecting and transporting thorny plants, bushes, shrubs, for preparing hedge and sometimes seed in threshing yards.

Hand hoe *(Khurpi):* This implement is a small one with a sharp edged triangular or sickle shaped blade fitted on a wooden handle operated in a sitting position. Several size and shapes are available. It is widely used for removing grass and other weeds from a field. It is pushed into the soil to dig the roots of weeds. Beating with the flat side also breaks small clods.

Cono weeder: This device, designed by Tamil Nadu Agricultural University, consists of two spike toothed wheels and a handle for pulling it. It is quite ideal to remove weeds from line-sown rice. The weeds are incorporated into the soil and ensure mild stirring of soil which increases aeration of soil and increases availability of nutrients.

Leveller: Levellers are used for levelling. This implement is used before wet sowing or transplanting of rice. All parts are made of wood. Its main part is a plank of about 1.5 m length, 37.5 cm height and 5 cm thickness. The lower edge of the plank that touches the soil is serrated. The implement is connected to the yoke with the help of two iron rings and fastened with ropes. It is operated by a person who keeps standing on the flat surface of the leveller.

Clod crusher: Clod crushers are used to break the clods after ploughing or harrowing. It consists of a plank of about 1.5 m length, and 10 cm width and 7.5cm thickness. It has got a beam too. The beam is connected to the yoke, and when the implement is dragged by the animals, the man stands on the plank and the clods get crushed. It also helps to compact and level the top soil.

Roller: Rollers are used widely in dry farming areas to compact the seedbed. In lawn making also rollers are used. It crushes clods and smoothens the surface soil. Wood, stone, or concrete are used for its construction. Animal drawn and tractor drawn rollers are available. Small sized rollers that can be drawn by human beings are also in use.

Wet land puddler: This implement is used to puddle the wet lands after the initial ploughings. It essentially consists of a number of blades fitted on a common axis. The axis rotates when drawn by the animals. These actions break the clods and mix the soil with water and the field gets puddled. Puddling destroys weeds, reduces the leaching of water, and softens the soil for transplanting.

Bund former: Bund former can be used to make bunds in a field, to make ridges for planting, and for making bunds along the contour as a soil conservation measure. The main part of the implement is a pair of mould boards fitted face to face. When the implement is pulled, soil from both sides is pushed towards the centre of the mould boards and a bund or ridge is formed.

Sickle: It is used mainly for cutting grasses, harvesting crops etc. There are different kinds of sickles like serrated or smooth but sharp cutting edges fixed in wooden handle.

Axe: It is made from iron. Its one end is sharp and other end is blunt. The blunt end has hole for fitting the wooden handle. Length of handle is 60-100 cm. It is used for cutting the woods, straw etc., for preparing hedge and for repairing the agricultural implements.

Rake: Two types of rakes are available (i) wooden rake (ii) iron rake. Four to eight prongs are provided on the headpiece of rake at equal distance. Prongs are pointed or wide depending upon use. It is used for mixing the manure into soil properly, to make small bunds in seedbed, to mix the seed with soil after broadcasting and useful in threshing yard.

Showel: It is prepared from iron square sheet. One end is sharp while opposite end is blunt which has a hole to fix the wooden handle. It is used for uprooting the seedlings from seedbed and digging and transporting the soil to a nearby place.

Wheel hoe: It is used for interculturing and weeding operations in the field crops. It is very useful particularly in light soils.

Dibbler: It is made from iron or wood. Its one end is pointed or sharp while other end is fixed with wooden handle. Used for dibbling of the seeds by making small holes in the soil.

Chisel plough: It is mainly used for breaking hard pans and for deep ploughing (60-70 cm) with fewer disturbances to the top layers. Its body is thin with replaceable cutting edge, have minimum disturbance to the top layers. It contains a replaceable share to shatter at the lower layers.

12

USE OF TILLAGE IMPLEMENTS

Reversible Plough

The reversible plough has two mouldboard ploughs mounted back-to-back, one turning to the right, the other to the left. While one is working the land, the other is carried upside-down in the air. At the end of each row, the paired ploughs are turned over, so the other can be used. This returns along the next furrow, again working the field in a consistent direction. In almost universal use on farms, they have right- and left-handed mouldboards enabling them to work up and down the same furrow. Reversible ploughs may either be mounted or semi-mounted and are heavier and more expensive than right-handed models, but they have the great advantage of leaving a level surface which makes seedbed preparation and harvesting easier. Very little marking out is necessary before ploughing can start and idle running on the headland is minimal compared with conventional ploughs.

Driving the tractor with the furrow-side wheels in the furrow bottom provides the most efficient line of draft between tractor and plough. It is also easier to steer the tractor and driving with the front wheel against the furrow wall will keep the front furrow at the correct width. This is less satisfactory when using a tractor with very wide front tyres, for although they make better use of the tractor power, the tyres may compact part of the last furrow slice turned on the previous run. The use of furrow widener or a longer mouldboard on the rear body will overcome the problem. The latter moves the soil further towards the ploughed land leaving more room for the tractor wheels on the next run.

Driving with all four wheels on unploughed land is another solution to the problem of wide tyres. Semi-mounted ploughs which can be hitched in are heavier and more expensive than right-handed models, but they have the great advantage of leaving a level surface which makes seedbed preparation and harvesting easier. Very little marking out is necessary before ploughing can start and idle running on the headland is minimal compared with conventional ploughs.

Semi-mounted ploughs can be hitched in a way that allows the tractor to run on unbroken land and pull the plough in correct alignment without any sideways movement (crabbing).

Seed cum fertilizer drill: Seed drills fitted with fertilizer dropping attachments are called seed-cum-fertilizer drills, They deliver both the seeds and fertilizers simultaneously in an acceptable pattern. Seed cum fertilizer drill has a large seed box which is divided lengthwise into two compartments, one for seed and another for fertilizers distribution. Seed drill may be classified as: (i) Bullock drawn (ii) Tractor drawn. Depending upon the method of metering the seeds, bullock drawn seed drill can be further divided into two groups viz; those in which seeds are dropped (a) by hand, (b) or mechanically. There are a number of bullock drawn implements which are used for sowing seeds in which seeds are dropped by hand. In different parts of the country it is made of different sizes and shapes.

Advantages

1. Seed and fertilizer can be sown at a time.
2. Seeds of different size can be sown.
3. The required quantity of seed and fertilizer can be sown.
4. The fertilizer can be sown below the seeds.
5. Inter row spacing can be adjusted according to our requirement.
6. It is made up of iron, hence its durability is more.

Thresher

Thresher is a machine to separate grains from the harvested crop and provide clean grain without much loss and damage. During threshing, grain loss in terms of broken grain, un-threshed grain, blown grain, spilled grain etc. should be minimum. Bureau of Indian Standards has specified that the total grain loss should not be more than 5 per cent, in which broken grain should be less than 2 per cent. Clean un-bruised grain fetch good price in the market as well as it has long storage life.

A mechanical thresher consists of i. feeding device (chute/tray/trough/hopper/conveyor) ii. threshing cylinder (hammers/spikes/rasp-bars/wire-loops/syndicator) iii. Concave (woven wire mesh/punched sheet/welded square bars) iv. blower/aspirator and v. sieve-shaker/straw-walker.

During threshing, the crop material is slightly pushed into the threshing cylinder through the feeding device, which gets into the working slit created between the circumference of the revolving drum having attached spikes and the upper casing. Simultaneously, the drum pulls the mass through the gap between the spikes and the upper casing with a varying speed. The angle iron ribs on the other hand, restrain the speed of the travelling of stalks clamped by the spikes. Due to this the spikes move in the working slit with a varying speed in relation to the shifting mass of material, which is simultaneously shifted, with a varying speed with respect to the upper casing. As a result, the material layer is struck several times by the spikes against the ribs, causing threshing of the major amount of grains and breaking stalks into pieces. As the material layer shifts towards the progressively converging slit of lower concave, its size reduces. The vibration amplitudes, therefore, decrease where as the speed of the layer increases. This causes mutual rubbing of the ear stalks, as well as rubbing of the ears against the edges of the concave bars and causes breaking of stalks depending on the concave clearance. Since the system is closed, the thicker stalk, which cannot be sieved through the concave, again joins the fresh stalk and the same process is repeated until the stalk size is reduced to the extent that it can pass through the concave apertures. Thus fine bruised straw is produced. The effective threshing process means that the loss of un-threshed kernels ejected with the straw through the concave and the loss of grain damage should be low and the amount of the material passed through the concave should be high.

Winnower

Winnowing is the process of separating grain from a mixture of grain, chaff and other impurities in an air stream from a natural or artificial source. The following devices are commonly adopted for winnowing process:

1. Winnowing basket
2. Winnowing fan
3. Winnower

Working of winnower: As the mixture falls into the air current, the lighter materials and chaff are blown through the opening for chaff and husk outlet by the air blast, while the good grains being heavier fall directly below and guided out by the grain outlet. Immature grains that are relatively lighter will be carried by the blast a little away from the good grain outlet and collected by the immature grain outlet. Purity adjustment mechanism is a wooden plank kept in between the grain and immature grain outlets in the air flowing path and is raised or lowered by a rack and pinion arrangement. It diverts major portion of the immature grains to the good grain outlet when lowered and immature grains are also collected; when raised the free flowing air will carry some quantity of good grains also to the immature grain outlet but improves the purity of grains in the good grain outlet. Proper rpm (revolution per minute) of the blower and feed rate and purity adjustments are essential for efficient winnowing of the grains from the other lighter fractions.

Plate 1. Different tillage implements

13

STUDY OF AGRO-CLIMATIC ZONES OF INDIA

Introduction

India exhibits a variety of land range and climatic conditions as evident from different types of soils and vegetation and showed a significant relationship among the soils, land, climate and vegetation. Reliable and systematic information on agro climatic regions based on homogeneity is the basic to maximize agricultural production and may help the country in planning and optimizing land use on sustainable basis.

Agro climatic zone is a land unit in terms of major climate and growing period which is climatically suitable for a certain image of crops and cultivars (FAO 1983). The agro-climatic classification is an extension of the climate classification keeping in view the suitability to agriculture. National Commission on Agriculture (1971) classified the country into 127 agro-climatic zones. The Planning Commission after examining the earlier studies on the regionalisation of the agricultural economy has recommended that agricultural planning be done on the basis of agroclimatic regions. During VII plan (1985-90) the country was divided into 15 broad agro-climatic zones based on the criteria of homogeneity in agro-characteristics such as rainfall, temperature, soil, topography, cropping and farming systems and water resources availability for effective land use in 1989. The object of present study is to delineate such regions as uniform as possible introspect of physiographic, climate, length of growing period (LPG) and soils for macro level and land use planning and effective transfer of agro-technology. The emphasis was given on the development of resources and their optimum utilization in a suitable manner within the frame work of resource constraints and potentials of each region.

Agro-Climatic Zones of India

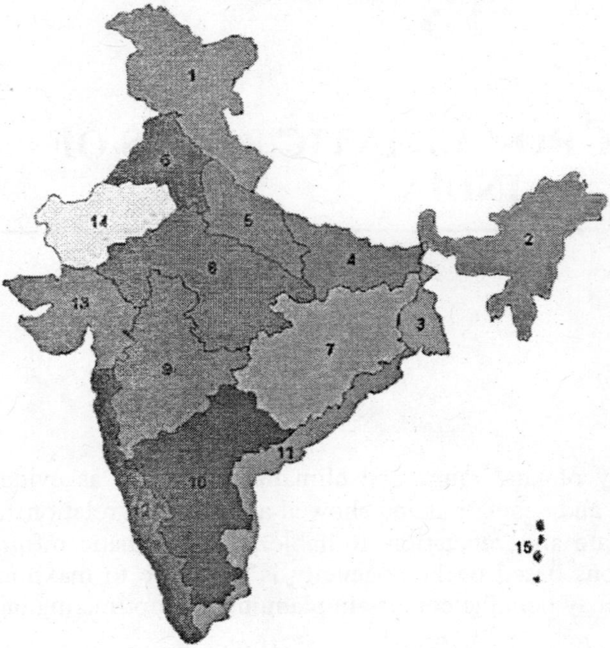

1. Western Himalayan
2. Eastern Himalayan
3. Lower Gangetic Plains
4. Middle Gangetic Plains
5. Upper Gangetic Plains
6. Trans Gangetic Plains
7. Eastern Plateau & Hills
8. Central Plateau & Hills
9. Western Plateau & Hills
10. Southern Plateau & Hills
11. East Coast Plains & Hills
12. West Coast Plains & Hills
13. Gujarat Plains & Hills
14. Western Dry Region
15. Islands

Fig. 1. Agro-climatic Zones of India

Table 1. Description of Agro climatic zone of the country:

Zone	Major Crops	Districts
Western Himalayan Region	The region consists of skeletal soils of cold region, podsolic soils, mountain meadow soils and hilly brown soils. Soils are generally silty loam with altitudinal variations. Lands have steep slopes in undulating terrain. Soils are prone to erosion hazards and slides and slips are quite common. Rice, maize, millets, wheat and barley are the main crops. Ginger, saffron, many temperature flowers and vegetables are grown in this region. This zone is having highest area (45.3%) under forests. Land use planning based on the concept that land up to 30% slope is suitable for agriculture on terraces, 30-50% slopes for horticulture and silvi-pastoral programmes, and above 50% slopes for forestry is a suggested strategy for development of the region.	Three distinct sub-zones of Jammu and Kashmir, Himachal Pradesh and Uttar Pradesh hills

Zone	Major Crops	Districts
Eastern Himalayan Region	High rainfall and high forest cover. Shifting cultivation (Jhum), practised in nearly one third of the cultivated area, has caused denudation and degradation of soils, with the resultant heavy runoff, massive soil erosion and floods in the lower reaches and basins. This area has a high potential for agriculture including forestry and horticulture.	Sikkim and Darjeeling hills, Arunachal Pradesh, Meghalaya, Nagaland, Manipur, Tripura, Mizoram, Assam and Jalpaiguri and coochbehar districts of West Bengal
Lower Gangetic Plains	This zone accounts for about 12% of the country's rice production. Soils mostly alluvial and are prone to floods and inundation of fields destroys standing crops. Sesamum, Jute, mustard, *rabi* maize and potato are emerging as new crops. The per capita land availability here is very low (0.095 ha). Scope for forage production and livestock rearing is very high.	The West Bengal – Lower Gangetic Plains region consists of four sub-regions
Middle Gangetic Plains	The rainfall is high and 30% of the gross cropped area is irrigated with cropping intensity of 142%. There is large area under salt affected (usar) lands. Zinc deficiency in rice is wide spread. It is suggested to put unculturable wasteland under silvi-pasture and culturable land under agro-forestry. Poultry, dairying and inland riverine fishery also should receive priority.	12 districts of eastern Uttar Pradesh and 27 districts of Bihar plains. Eastern U.P. has been further sub-divided into nine regions based on the heterogeneity in soil, land use, topography and climatic factors.
Upper Gangetic Plains	The zone has 144% cropping intensity. Irrigation is largely through canals and tube wells with good potential for exploitation of ground water. In all the Diara lands (flood prone areas) development of fruit trees is important. Genetic improvement through cross breeding and increasing the area under fodder crops is important.	32 districts of Uttar Pradesh divided into three sub-zones of Central, North-West and South –West U.P
Trans-Gangetic Plains	Rice-wheat system is prevalent. The major features of the area are: highest net sown area, highest irrigated area, least poverty level, high cropping intensity (170%) and high ground water utilization. There is need to evolve short duration genotypes and also to diversify of the cropping.	Punjab and Haryana, Union Territories of Delhi and Chandigarh and Sriganga-nagar district of Rajasthan It is delineated into three sub-zones, namely, foothills of Shiwalik and the Himalayas, plains (Semi arid) and arid zone bordering the Thar desert.

Zone	Major Crops	Districts
Eastern Plateau and Hills	The soils of the region are shallow and medium in depth and the topography is undulating with a slope of 1 to 10%. Rainfall is nearly 1300 mm. Integrated water shed development approach to conserve soil and rainwater should be strengthened. Irrigation is largely through tanks and tube wells In kharif, 82% of the area is under rice. Most soils are acidic and in some areas application of lime is necessary. Cultivation of crops like redgram, groundnut, and soybean in uplands is to be encouraged. Mustard and vegetables are to be grown in irrigated areas. The rehabilitation of degraded peripheral forests is to be taken up on a large scale.	The eastern Plateau and Hills region consists of the following sub-regions: i. Sub-region of Wainganga, Madhya Pradesh Eastern Hills and Orissa inland, ii. Orissa Northern and M.P. Eastern Hills and plateau. iii. Chotanagpur North and Eastern Hills and plateau. iv. Chotanagpur South and West Bengal Hills and Plateau, and v. Chattisgarh and South-Western Orissa Hills.
Central Plateau and Hills	Irrigation and intensity of cropping are low. Per capita availability of land is very high (0.446 ha). Since 75% of the area is rainfed, a watershed management programme is to be implemented. Food crops should be replaced by oil seeds.	This zone comprises of 46 districts of Madhya Pradesh, Uttar Pradesh and Rajasthan.
Western Plateau and Hills	This region forms a major part of peninsular India, with an annual average rainfall of 904 mm. Net sown area is 65% and only 12.4% area is irrigated. Sorghum and cotton are the major crops. This zone is known for the best quality oranges, grapes and bananas. Farmers are adopting sprinklers and the drip methods of irrigation, particularly, for fruit and vegetable crops. 11% forest.	This zone comprises of major parts of Maharashtra, parts of MP and one district of Rajasthan and is divided into four sub-zones.
Southern Plateau and Hills	Typically semi arid zone. Rainfed farming is adopted in 81% of the area and the cropping intensity is 111%. Low value cereals and minor millets predominate in the cropping systems. The adoption of proven dryland technology in the watershed areas should aid agriculture in this area. Crop diversification has to be intensified and crops that require less moisture should be preferred. Poultry has developed quickly in many areas of the zone.	This zone comprises of 35 districts of Andhra Pradesh, Karnataka and Tamil Nadu.

Zone	Major Crops	Districts
East Coast Plains and Hills	Rice and groundnut are the important crops as the soils are alluvial. Nearly 70% of the cultivated area does not have irrigation facility. Tanks account for nearly 20% of the irrigated area in the zone. Desilting tanks, strengthening of bunds and structures and improvement of field channels programmes need to be taken up. Drainage, particularly in the south coastal Andhra Pradesh (Krishna – Godavari delta) and Cauvery delta areas id a vital need. Alkaline-saline soils in the region total up to 4.9 lakh ha. Area under waste lands estimated to 25.33 lakh ha. The zone with over 2,000 km of coastline. Brackish water fisheries and aquaculture hold great promise in this area.	This zone consists of six sub-zones (i) Orissa coastal, (ii) North coastal Andhra and Ganjam, (iii) South coastal Andhra, (iv) North coastal Tamil Nadu, (v) Thanjavur and (vi) South coastal Tamil Nadu.
West Coast Plains and Ghats	This zone has a variety of crop patterns, rainfall and soil types. Major area under plantation crops and spices and fisheries. Cropping intensity is 124%. Productivity of rice and millets is low and there is need for diversification to horticulture crops. The approach of homestead (group farming) system (one of the agro-forestry systems) of reclaiming and using khar lands (saline soils) or pokhali lands (acidic soils) needs to be planned and implemented. This zone is important for multi-storied cropping.	This zone runs along the west coast, covering parts of Tamil Nadu, Kerala, Karnataka, Maharastra and Goa
Gujarat Plains and Hills	The zone is arid with low rainfall, 22.5% of the area is irrigated, irrigation largely through wells and tube wells. Only 50% of the cultivated area is under food crops resulting in food deficit. However, it is an important oilseed zone. The cropping intensity is 114% and nearly 60% of the zone is considered drought prone. The major thrust should be on rainwater harvesting, dry farming and canal and ground water management. The long coastline and river deltas should be used fully for developing marine fishing and brackish/backwater aquaculture.	This zone consists of 19 districts of Gujarat classified into seven sub-zones

Zone	Major Crops	Districts
Western Dry Region	This region comprises of nine districts of Rajasthan and is characterized by hot sandy desert, erratic rainfall, high evaporation, no perennial rivers and scanty vegetation. The ground water is deep and often brackish. Famine and drought are common features. The average annual rainfall is only 395 mm with wide fluctuations from year to year. The forest area is only 1.2%. The land under pastures is also low (4.3%). The cultural waste and fallow lands are substantial, accounting for nearly 42% of the geographical area. The net irrigated area is only 6.3% of the net sown area. Cropping intensity is 105%. Pearl millet, cluster bean (guar) and moth are the lead crops in kharif and wheat and gram in rabi, but the yield levels per hectare are low. Increasing tree cover is important to (a) check desertification, (b) provide fodder to livestock, (c) meet the fuel needs of the population, and (d) provide timber implements.	This region comprises of nine districts of Rajasthan
Islands region	This zone is typically equatorial. Rainfall of 3,000 mm is spread over eight to nine months. It is largely a forest zone having large undulating areas leading to heavy loss of soil due to runoff. Nearly half of the cropped area is under coconut. This is the smallest zone.	Island territories of the Andaman & Nicobar and Lakshadweep

■■■■

14

IDENTIFICATION OF COMMON WEEDS

INTRODUCTION

Weeds are defined as the plants growing out of place and out of time. An enormous loss in crop yields usually 50-80% with inferior quality of produce is due to the presence of weeds. Weeds are found in the field throughout the year in different seasons under different habitats. They may be of different families and growth habits having different modes of reproduction, dispersal and propagation. Development or selection of an appropriate and effective weed management program is decided on the sound knowledge of weed identification (biology). The weed biology is related to their geographic distribution, habitat, growth, population dynamics and communities. The objective of identification is to acquire broad knowledge about weed species and their intensity to estimate the magnitude of their harmful effects on crop production and planning an effective weed management strategy. If one can accurately identify a plant that is causing problems in concerned landscape, means the first step toward a good management plan. The weeds found in the field can be broadly divided into following categories:

(A) Growth duration (life cycle, ontogeny): Annual, Biennial and Perennial weeds.

(B) Morphological: Grasses, Sedges and Broadleafs.

(C) Nature: Associated, Crop bound, Parasitic, Facultative and Obligate weeds

The weeds can be identified on the basis of following features in the field.

1. Leaf: Colour, size, shape, arrangement
2. Stem: Colour, size, nature (erect, prostrate, angular trailing etc.) nodes and internodes, solid, hollow, woody, tender
3. Branches : Branched, unbranched, branching pattern and arrangement
4. Flowers : Colour, size, type of inflorescence
5. Roots parts: Shallow, deep, tap root, adventitious

6. Underground parts: Rhizomes, nuts, bulb etc.
7. Juncture points: The colour, shape, size of any plant part like hair, ligule, auricle, glands etc. present at the joining point of stem with root, leaf with stem, inflorescence with the main stem/branch etc.
8. Fruits/seeds : Colour, size, shape and type.
9. Special points: Nature of plant sap (milky, juicy, gum etc.) and its colour, special modification on the plant etc.

Materials required

Manual on weed identification, Books on Botany, Taxonomy, Weed Science, Herbaria, HB Pencil, White paper etc.

Procedure

Visit and record the weed species present in different farming situations such as up lands, low lands and irrigated fields along with waste lands and farm ponds. Further, note down the salient characteristics after locating their types, study their habitat, morphology, growing season and mode of propagation. The help may be taken from books, manuals, herbaria, exhibits and other references for correct and scientific identification of weed specimen.

Observation

As you get a weed plant, make a visual observation, understand its basic characteristics and write their common name, botanical name, category, family and any special characteristic which helps in their effective control. The common weeds found in the country are presented with their common and botanical names in tabular form (Table 1 and 2) and the pictorial presentation of weeds is given in Plates 1-10.

Table 1. List of common grassy and narrow leaf weeds

Grasses:

Scientific name	Common name	English name
Agropyron repens	-	Quack grass
Andropogan leniger	Boor grass	-
Aristida depressa	Lampra grass	Three own
Cenchrus ciliaris	Anjan	Buffel grass / African foxtail
Cenchrus purea	White anjan	-
Cenchrus setigerus	Dhaman Black	-
Cenchrus biflorus	Bhroorat/baroot	Sand bur

Scientific name	Common name	English name
Cynodon dactylon (L.) Pers.	Doob	Bermuda grass
Cyperus iria	Dachab	Umbrella sedge
Cyperus esculantus	Motha	Yellow nutsedge
Cyperus rotundus	Motha	Purple nutsedge
Dactyloctenium aegyptium	Makra grass	Egyptian crowfoot grass
Desmosta chyabipinnata	Dab/kusha	-
Digitaria sanguanalis	Jhernia grass	Crab grass
Echinochloa colonum	Jangli dhan	Jungle rice
Echinochloa crusgalli	Sawan/ kodon	Barnyard grass
Eleusine indica	Mandla/ Balrara	Goose grass
Eragrostis poacides	Chidi grass/ lover grass	Lover grass/ stink grass
Eremopogon foveolatus	Dab	-
Imperata cylindrica	Thatch grass	Congo grass
Lesiurus sindicus L.	Sewan grass	Desert grass
Oryza sativa L.	Jangli rice	Wild rice
Paspalum distichum		Knot grass
Phalaris minor Retz.	Guillidanda/ gehunka mama	Little seed canary grass
Poa annua L.	Sua grass	Annual blue grass/ sweet grass
Saccharum munja L.	Moonj/ kuncha	-
Saccharum spontaneum L.	Kans	Wild Cane/ thatch grass
Setaria glauca	Bandra-bandri	Yellow foxtail
Sorghum halepense	Barru/Sudan grass	Johnson Grass

Table 2. List of common broad leaf weeds

Scientific name	English name	Scientific name	English name
Abutilon indicum	Country Mallow	*Eichornia crassipes*	Water hyacinth
Acalipha indica	Indian acalypha	*Euphorbia hirta*	Phyician's spurge
Achyranthes aspera	Pricklychaff Flower	*Gomphrena celosioides*	Gomphrena
Ageratum conyzoides	Goat weed	*Ipomoea carnea*	Silver Morning Glory

Scientific name	English name	Scientific name	English name
Agropyron repens	Quack grass	*Lantana camera*	Wild sage
Amaranthus spinosus	Spiny amaranth	*Leucas aspera*	Thumbai
Amaranthus viridis	Parrot amaranthus	*Melilotus alba*	White sweet clover
Ammania baccifera	Fire feaf	*Melilotus indica*	Yellow sweet clover
Argemone mexicana	Mexican Poppy	*Opuntia dellenii*	Prickly pear
Boerhavia diffusa	Spreading hog weed	*Orobanche cernua*	Broom rape
Borreria hispida	Buttonweed	*Oxalis corniculata*	Lady's sorrel
Calotropis gigantea	Giant swallow wort	*Parthenium hysteroporus*	Carrot grass
Carthamus oxycantha	Wild safflower	*Paspalum distichum*	knot grass
Cuscuta chinensis	Dodder	*Phyllanthus niruri*	Stonebreaker
Cassia tora	Sickle Senna	*Physalis minima*	Ground cherry
Celosia argentia	White cock's comb	*Pistia stratiotes*	Water lettuce
Chenopodium album	Lamb's quarter	*Portulaca oleracea*	Indian purslane
Cichorium intybus	Blue daisy	*Rumex acetosella*	*Red sorel*
Circium arvense	Canada thistle	*Setaria glauca*	Yellow foxtail
Cleome viscose	Dog mustard	*Setaria viridis*	Green foxtail
Commelina benghalensis	Day flower	*Sonchus arvensis*	Perennial Sowthistle
Convolvulus arvensis	Field bind weed	*Striga asiatica*	Witch weed
Coronopus didymus	Swine cress	*Tephrosia purpurea*	Wild indigo
Datura stramonium	Jimson weed	*Tribulus terrestris*	Small caltrop
Digera arvensis	False Amaranth	*Tridax procumbens*	Mexican daisy
Eclipta alba	Eclipta/Bringraj	*Xanthium strumarium*	Cocklebur

Phalaris minor

Avena ludoviciano

Polypogon mospliensis

Poa annua

Lolium rigidum

Chenopodium album

Coronopus didymus

Rumex dentatas

Fumaria parviflora

Convoluvulus arvensis

Chenopodium muerale

Asphodelus tenuifloius

Plate 1 Photographs of Rabi crops

Vicia sativa

Melilotus indica

Medicago denticulata

Cichorium intybus

Anagallis arvensis

Malwa parviflora

Stellaria media

Rumex spiniosus

Argemone maxicana

Lathyrus aphaca

Cirsium arvense

Pluchea lanceolata

Plate 2 Photographs of Rabi crops

Trianthema portulacastrum

Digera arvensis

Physallis minima

Euphorbia hirta

Corchorus tridems

Celosis argentia

Crotolaria medicaginea

Cleome viscosa

Mollugo cerviana

Cenchrus ciliaris

Cyperus rotundus

Cenchrus longispinosus

Plate 3 Photographs of Kharif crops

Dactyloctenium aegypticum

Brachiaria reptans

Eleusine indica

Bulbosyllis barbata

Digitaria sanguinalis

Amaranthus viridis

Tribulus terrestris

Amaranthus spinosus

Phyllanthus niruri

Ipomoea pestigridis

Corchorus olitorius

Ipomoea lacunosa

Plate 4 Photographs of Kharif crops

Echinochloa colona

Echinochloa glabrescens

Leptochloa chinensis

Ischaemum rugosum

Eleusine indica

Cyperus difformis

Eleocharis

Cyperus rotundus

Cyperus iria

Scirpus supinus

Paspalum distichum

Scripus maritimus

Plate 5 Photographs of Grassy weeds and Sedges in rice crops

Sphenoclea zeylenica

Ammania baccifera

Caesulia axillaris

Eclipta alba

Ipomoea aquatica

Lemna minor

Ludwigia parviflora

Marsilea quaderifolia

Saggitaria guanensis

Commelina benghlensis

Monochoria vaginallis

Mazus spp.

Plate 6 Photographs of Broadleaf weeds of rice crops

Desmostachya bipinnata

Imperata cylindrica

Cynodon dactylon

Achyranthus aspera

Lantana camara

Abutilon indicum

Croton sparsiflorus

Calotropis procera

Ipomoea carnea

Sorghum halepense

Ageratum conzyoides

Suaeda fruticosa

Plate 7 Photographs of Perennial weeds crops

Parthenium hysterophorus

Cannabis sativa

Xanthium strumarium

Helianthus annus

Carthamus oxycantha

Cirsium arvense

Boerhavia diffusa

Argemone maxicana

Datura stramonium

Cassia occidentalia

Cassia tora

Sida acuta

Plate 8 Photographs of Noncropped area weeds crops

Eichhornia crassipes

Typha angustata

Salvinia molesta

Potamogeton nodosus

Hydrilla verticillata

Nymphea odrata

Nelumbo lutea

Vallisneria spiralis

Trapa natans

Pistia stratiotes

Najas minor

Alterneenthra philoxeroides

Plate 9 Photographs of Aquatic weeds crops

Inflorescence of Orobanche

Orobanche aegyptiaca infestation in mustard

Cuscuta chinensis infestation on clordendron and lucern

Loranthus infestation on tree

Loranthus

Striga asiatica

Striga asiatica flower and seeds

Plate 10. Photographs of Parasitic weeds crops

15

METHODS OF HERBICIDE APPLICATION

INTRODUCTION

Herbicides can be applied on the crop or a field in a variety of ways to control weeds either as single component products or as combination of products by considering their properties and weed spectrum without harming the crop plant. Different plants species respond differently to same herbicide and same plant species respond differently to different herbicides. This is a foundation for phenomenal achievement in modern chemical weed management where objective is to kill weeds and retain others at the same time at place. An improper application results in either cause poor weed control or severe crop injury. The main objective to study methods of herbicides application is given below :

- To provide early season weed control in favour of crop.
- To decrease the weeding cost, labour use and to save time.
- To ensure the effective use of herbicide.
- To protect the non target and economical vegetation from the herbicide without any injury.
- To achieve the higher weed control efficiency

Methods of Herbicide applications

Herbicides differ widely in their physical, chemical and biological properties. They are selective or non-selective to crops and have wide or narrow spectrum in weed control. Herbicides within a chemical family are structurally related and have essentially the same mode of action in plants with a few notable exceptions. Similarly, herbicides within a chemical family also often vary in selectivity, a result of physico-chemical differences that cause them to behave differently in the soil or plant system. Methods of application of herbicides are decided largely by their modes of action and selectivity.

Herbicides are mostly soil or foliage applied to cropped and non cropped area. When applied to the soil, they are known as soil applied herbicides and when applied to the foliage of the weed, referred as foliage applied herbicides.

Most of pre plant incorporation and pre emergence herbicides are soil applied while post emergence herbicides are applied on foliage in cropped areas or on waste lands.

(1) Soil application of herbicides

(i) Surface application: They are commonly applied on soil surface where they are left as such (undisturbed) or incorporated into the soil. Even when left on the soil surface, applied herbicide should be able to move into upper 3-4 cm of soil under the influence of soil moisture either through rain or irrigation to kill the germinating weeds. The triazines, ureas and anilides herbicides are few examples. The volatile herbicides like trifluralin and fluchloralin are incorporated into soil physically by light ploughing or with light irrigation to the depth of 5 cm to prevent their rapid losses by volatilization or photodecomposition. The optimum depth of incorporation of different herbicides for achieving selective control of weeds varies considerably. For example EPTC is incorporated 5 to 7.5 cm deep in soil whereas trifluralin and nitralin are incorporated not beyond 5 cm depth. On the other hand diallate and triallate herbicides are mixed with soil only upto 2.5 cm depth.

(ii) Subsurface layering: Herbicides are applied in concentrated band about 7-10 cm below the soil surface to inhibit the growth of new shoots of perennial weeds such as *Cyperus rotundous* and *Convolvulus arvensis*. The subsurface application is done through the special flood nozzles under the cover of sweep hoods introduced below the soil.

(iii) Broadcast and band application: Broadcast application of herbicide is made over an entire area without leaving and intentional gaps whereas, band application usually means treating a narrow strip (restricted area) directly over or on the crop row usually 30-35 cm. The band application is cost saving as it reduces the quantity of herbicides required. In band method of application, the inter rows are cultivated manually later on to remove weeds from the untreated areas. For example a 30 cm wide band of a herbicide applied over crop rows that were spaced 90 cm apart, will require one third of the quantity of herbicides needed for its application as compared to broadcast method. This will also help in the control of perennial and some other weeds, which may be resistant to the herbicide.

(iv) Soil fumigation method: Depending upon the nature of the soil fumigant herbicides can be applied either by-

a. soil injection (e.g chloropicrin),

b. releasing under sealed plastic covers (e.g methyl bromide) and

c. soil surface application (e.g metham)

(2) Foliage application of herbicides

(i) Blanket application or over to top application - The herbicides are applied uniformly to the standing crop irrespective of the location of crop plants.

The highly selective herbicides e.g. 2, 4-D in wheat, MCPB in pea, 2,4-DB in lucerne and picloram in sugarcane are preferred for blanket application.

(ii) **Directed spraying:** The application of herbicides is directed to weeds growing in inter rows of the widely spaced tall crops by avoiding the crop foliage as much as possible (e.g. paraquat in sugarcane). Directed spraying is done by changing the setting of nozzles or fitting the nozzle in the hood or shield to intersect the base of the crop plants just above the soil line. This saves the crop plants from herbicide injury and increases weed control. Usually the selective herbicides are used in crops but in hardy crops with greater plant height over weeds, the non selective herbicides can be applied by using nozzle attached with the hood. One important prerequisite of directed spraying is that farmer must get his crop well up over the weeds at the time of treatment by modification in agricultural practices like good seedbed preparation and early planting. These modifications help the crop plants to rapid emergence and gain a height advantage over the weeds. Orchards and plantation are good venues of directed herbicide sprays.

(iii) Protected spraying: Non-selective herbicides can be applied by covering the crop plants with plastic or metallic covers to obtain selective weed control in distantly spaced (Widely sown) vegetables and ornamental plants before application of the herbicide. This can be done in 3-4 rows at a time and shifting the covers to the untreated rows each time. This method is very laborious but it pays itself when weeding of high value vegetable crops and ornamental plants are involved.

(iv) Spot treatment: Spot treatment is usually done on small areas of serious weed infestation to kill it and to prevent its spread by leaving the weed free gaps untreated. Noxious perennial weeds are effectively controlled by spot treatment with potent herbicides.

Other methods of herbicides application

(a) Direct contact application (DCA): It involves all the techniques such as wiping, rubbing and smearing of herbicides on the target plant surface. It is done by using herbicide laden wax bar, herbicide cloth mulch, roughing gloves or hand brush. The herbicide laden wax bars are dragged against weeds growing much taller to the crop plants. Herbicide cloth mulches are placed in the crop inter rows. The roughing gloves carry arrangement to smear herbicide on the weed gripped by the worker.

(b) Soil injection: Herbicides like ethylene dibromide, carbon disulphides and vernolate are applied at prescribed spacing before planting of crops by soil injection.

Method of treating brush and tree weeds

(a) Basal bark treatment – The basal bark of the brush or tree to the length of 30 cm is peeled and herbicides are applied to the peeled portions only.

(b) Cut stump treatment - The herbicide application is made on the cut surface after sawing the tree just above the ground level.

(c) Frill, knotch and injection method - The frills and knotches are made with sharp tool into the tree weed at convenient height. The frill and notches are filled with herbicide. The herbicides are also injected into the holes made in the tree trunk up to 2.5 cm thickness. These methods are commonly adopted on thick stem trees of 8 cm or more in diameter.

Girdling and frilling are methods of killing standing trees that may be done with or without an herbicide. However, the groove must completely encircle the trunk. Frilling is a variation of girdling in which a series of downward angled cuts are made completely around the tree, leaving the partially severed bark and wood anchored at the base

Precautions during application

- The spray tank should be at least one third full with clean water before addition of any concentrate for proper mixing of concentrate. Then concentrate is being put in slowly.
- Wettable powder formulation should be made up into a paste and then diluted before adding to the spray tank.
- Fill up the tank up to two third of its volume and left one third empty for proper spraying.
- Pumping should stop as soon as the tank is empty and a mark should be left in the field to indicate to start the spraying again.
- It is worth full to divide the whole area to be treated into convenient blocks for uniform spraying for pre emergence applications.
- Early morning is considered ideal period for herbicide application as the wind is very gentle during that period.
- Spraying with desired spray width is highly required. Spraying along the rows can facilitate easy swath matching.
- Avoid spraying foliage active herbicides if rains are expected in next two hours or so. However, a light rain or irrigation is often beneficial for a soil active herbicide.

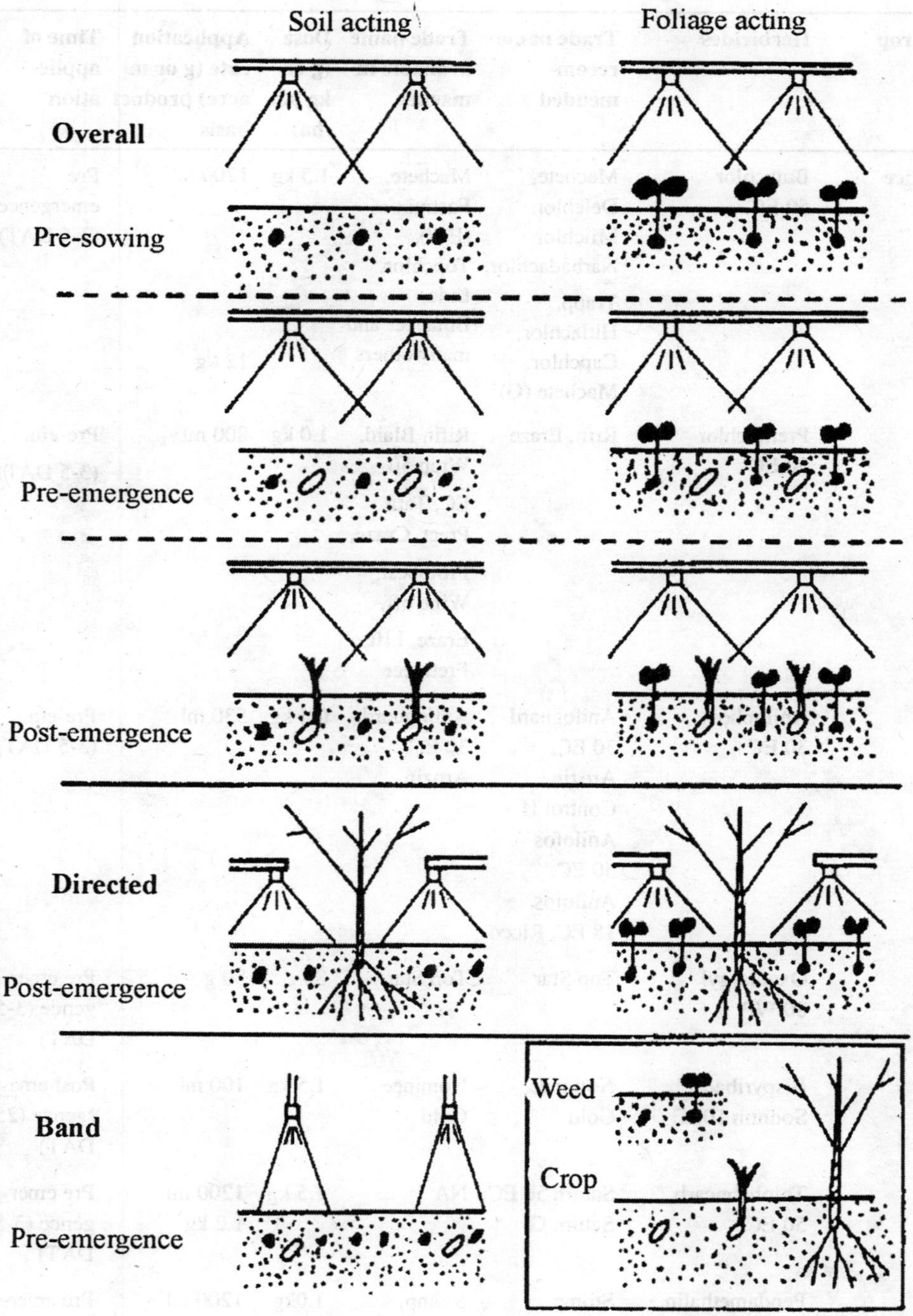

Fig. 1. Pattern and response of herbicide application

Table 1. Herbicides recommended for weed control in different crops

Crop	Herbicides	Trade name recom- mended	Trade name available in market	Dose (g or kg a.i. /ha)	Application rate (g or ml acre) product basis	Time of applic- ation
Rice	Butachlor 50 EC	Machete, Delchlor, Milchlor, Narbadachlor, Trapp, Hitlachlor, Capchlor, Machete (G)	Machete, Fastmix (EW), Topchlor, Lido, Butaveer and many others	1.5 kg	1200 ml 12 kg	Pre emergence (3-5 DAT)
	Pretilachlor 50 EC	Rifit, Eraze	Rifit, Blaid, Wrap 50 EC, Tata Preet, Craze, Protobest, Wipe off, Eraze, Lift, Pretiveer	1.0 kg	800 ml	Pre-em. (3-5 DAT)
	Anilophos 30 EC	Aniloguard 30 EC, Arozin, Control H, Anilofos 50 EC, Anilofos 18 EC, Ricco	Aniloguard 30 EC, Arozin	0.4 kg	530 ml	Pre-em. (3-5 DAT)
	Oxadiargyl 80 WP	Top Star	Top Star	80 g	50 g	Pre emer- gence (3-5 DAT)
	Bispyribac- Sodium 10EC	Nominee Gold	Nominee Gold	1.5 kg	100 ml	Post eme- rgence (25 DAT)
	Thiobebncarb 50 EC	Saturn 50 EC, Saturn G	NA	1.5 kg	1200 ml 1.2 kg	Pre emer- gence (3-5 DAT)
	Pendamethalin 30 EC	Stomp	Stomp, Penditera, Pendistar, Panida, Dost	1.0kg	1200 ml	Pre emer- gence (3-5 DAT)

Crop	Herbicides	Trade name recom-mended	Trade name available in market	Dose (g or kg a.i. /ha)	Application rate (g or ml acre) product basis	Time of applic-ation
	Flucholralin 45 EC	Basalin 45 EC Basalin g	Basalin	0.54-0.67 kg	500-600 ml	Pre emerge-nce (3-5 DAT)
	Metsulfuron (10%) + Chlorimuron (10 %) RM + (0.2% Surfactant)	Almix 20 WP	Almix	4 g	8 gm	20-25 DAS
	Ethoxysulfuraon 15 WDG	Sun Rice	Sun Rice	15g	50 g	20-25 DAS
	2,4 D Ester 36-38 EC	Eco Weed 38 EC, Knockweed, Weedmar	Eco Weed 38 EC, Knock-weed, Weedmar	580 ml	500 ml	20-25 DAS
	2,4 D Amine 58 % EC	Weedor	Weedor, Zura, Champio	580 ml	400 ml	20-25 DAS
Bajra	Atrazine 50 WP	-	Atrataf, Solaro, Atraking	0.25 kg	200 gm	1-10 DAS
Maize	Atrazine 50 WP	-	Atrataf, Solaro, Atraking	0.75 kg	600 gm	1-10 DAS
Sorghum	Atrazine 50 WP	-	Atrataf, Solaro, Atraking	0.50 kg	400 gm	1-10 DAS
Sugar-cane	Atrazine 50 WP	-	Atrataf, Solaro, Atraking	1.25 kg	1000 gm	1-10 DAS or after 1st Irrigation
	2,4 D Ester 36-38 EC	Eco Weed 38 EC, Knock-weed, Weedmar	Eco Weed 38 EC, Knock-weed, Weedmar	580 ml	500 ml	50-90 DAS
	2,4 D Amine 58% EC	Weedor	Weedor, Zura, Champio	580 ml	1000 ml	50-90 DAS
	2,4 D Na salt 80% WP	Weedor	Safaya	580 ml	1000 ml	50-90 DAS
Cotton	Trifluralin 48 EC	Trelan	Trilex, Toofan	1.0kg	800 ml	PPI and Post emergence

Crop	Herbicides	Trade name recom-mended	Trade name available in market	Dose (g or kg a.i. /ha)	Application rate (g or ml acre) product basis	Time of applic-ation
	Pendamethalin 30 EC	Stomp	Stomp, Penditera, Pendistar, Panida, Dost	1.0kg	1250 ml	Pre emergence (3-5 DAT)
Wheat	Clodinafop 15 WP	Topik, point, Moolah, Rakshak Plus, Jai Vijay	Topik, Point, Moolah, Rakshak Plus, Jai Vijay, Jhataka, Sartaj, D-Fop, Topple, Libra,	60 g	160g	30-45 DAS
	Pinoxaden 5 EC	Axial	Axial	50	400 ml	30-45 DAS
	Sulfosulfuron 75 WP	Leader, Safal-75, SF-10	Leader, Safal-75, SF-10, Wazir, Sonihal, Fateh, Nikhar, Munto, Teacher, Kaiser	25 g	13g+ surfactant	30-35 DAS
	Meso (3%) + Iodosulfuron (0.6%)	Atalntis	Atlantis	12+2.4	160 g+S	30-35DAS
	Isoproturon 75 WP	Nocilon, Iso-gaurd, Milron, Mohlan, Tol-kan, Arelon	Isogaurd, Isover, Nocilon, Gilron	0.5 kg	500 gm	30-35DAS
	Fenoxaprop 10 EC	Puma power	Puma Power	100g	100g+S	30-45DAS
	Metsulfuron Methyl 20 WP	Algrip	Algrip, Sumo, Convo, Sinmet	4 g	8g +S	30-35DAS
	Carfentazone 40 WDG	Affinity	Affinity	20	20G	30-35DAS
	Pendamethalin 30 EC	Stomp	Stomp, Penditera, Pendistar, Panida, Dost	1.0kg	1250 ml	Pre emergence (3-5 DAT)

Crop	Herbicides	Trade name recom-mended	Trade name available in market	Dose (g or kg a.i. /ha)	Application rate (g or ml acre) product basis	Time of applic-ation
	2,4 D Ester 36-38 EC	Eco Weed 38 EC, Knockweed, Weedmar	Eco Weed 38 EC, Knock-weed, Weedmar	580 ml	500 ml	50-90 DAS
	2,4 D Amine 58 % EC	Weedor	Weedor, Zura, Champio	580 ml	1000 ml	50-90 DAS
	2,4 D Na salt 80% WP	Weedor	Safaya	580 ml	1000 ml	50-90 DAS
Non cropped areas	Glyphosate 58 % SL	Round-up, Glycel	Roundup, Glyscel, Roundoff, Popular, Weedol	1.5-2.0 %	1.5-2.0%	Post emergence
	Paraquate 24SL	Gramxone	Gramoxone, Alliquit	0.5 % product basis	0.5%	Post emergence
Vege-tables	Pendamethalin 30 EC	Stomp	Stomp, Penditera, Pendistar, Panida, Dost	1.0kg	1250 ml	Pre emergence (3-5 DAT)
	Trifluralin 48 EC	Trelan	Trilex, Toofan	1.0kg	800 ml	PPI and Post emergence
	Oxyflurofen 23.5% EC	Gaol 23.5 % EC	Oxygold, Ronaldo, goal	0.15-0.2 kg	200 ml	Pre and post Emergence

■■■■

16

NUMERICAL EXERCISES ON HERBICIDES

INTRODUCTION

The selection of herbicides largely depends on soil types, cropping situation, cropping pattern, presence of weed species and crop status for effective control of weeds in prevailing environment. The commercial products (herbicides) contain active ingredient (Chemical responsible for herbicidal effect), adjuvant and other inert material as filler depending upon types of formulations. Therefore, it is necessary to recommend the dose of herbicide based on active ingredients for effective control. Sometimes, the dose (rate) of herbicide also varies with intensity of weed infestation, soil types, climate, crop tolerance, management practices and method and time of application. The calculation of spray volume is required to dissolve the required quantity of formulation for uniform distribution and equal effectiveness. The over or under dosing of herbicide can make a marked difference between the success and failure of weed control programmes. Sometime over dose of herbicide may change the property of herbicide like a selective herbicide becomes non selective in nature and may cause considerable damage to economic plants.

Rate of application: It is the amount of active ingredient or acid equivalent of herbicide applied to a unit area of land or water body. It is usually given in terms of kg ai/ha. Or a.e/ha.

Active ingredient (a.i.): A chemical in commercial product that is directly responsible for its herbicidal effect is called active ingredient. The ingredient may also be given as per cent by weight or volume eg. herbicide concentrate 40% w/v contains 400 g of active ingredient per litre of the liquid product.

Acid equivalent (a.e): Refers to that part of a formulation which theoretically can be converted to the acid. Some herbicide structures are active organic acids and some are not dissolved in water (eg. phenoxyalkanoic, picloram, chloramben, cacodylic acid). They are prepared in the form of their salts and esters for ease in their field application. For instance 2, 4-D in acid form is water insoluble then we have to use its sodium and amine salts and esters. The acid equivalent (a.e) of sodium salt of 2,4-D is 92.5%. It indicates that 2,4-D dichloro phenoxy acetic acid is 92.5% in sodium salt of 2,4-D. Both a. e. and a. i.

are generally expressed in kilogram (1000g) per hectare (10,000 sq. m.) for either dry or liquid formulations. Liquid formulations may indicate both per cent active ingredient and acid equivalent on weight per liter. In such cases, the concentration in terms of acid equivalent may be considered.

Career (Diluents): It is solid or liquid material used to increase the volume of a chemical compound for their uniform application over the target area. The universal career of herbicide is water because of its negligible cost and freedom from interactions. Sometimes career oils are also used for a quick knock down effect of weeds in non-cropped area because they have low surface tension (40.0), creep on the sprayed vegetation and enter into the crowns and rosettes of the grasses and sedges. Water with its high surface tension (75.0) proves superior on the dense growth of weed. Hard water should not be used for spraying as it may precipitate the herbicides. In recent years air has been used as career. Motorized ground or aerial sprayers are equipped with specially designed exhaust nozzles. The air atomizes the liquid concentrate into 30-150 micron diameter droplets and ejects them out of the nozzle. These are easily prone to drift by air but this can save labour, time and spraying cost.

Volume: It refers to total quantity of liquid applied per unit area. The quantity of toxicant reaching the target weed or soil is as much as important as spray volume. The spray volumes for herbicides have been classified into 5 classes from high volume to ultra low volume sprays (Table 1). The high volume spray provides thorough coverage of target plants to the point of runoff or drip eg: contact herbicides.

Common name: Technically accepted short form of the chemical in herbicide is called common name. Common names are agreed upon by a committee on terminologies.

Chemical name: It gives the full molecular structure of herbicide chemical according to IUPAC rules.

Trade name: Trade name is the name offered by the manufacturer.

- These 3 names are approved by different organizations or institutes which are considered authority on this subject

The following formula used for calculating the quantity of formulation required, quantity of herbicides in a.i. and amount of water to dissolve the required quantity of formulation are given below:

Quantity of commercial formulations (herbicide) required (kg/ha)

$$= \frac{\text{Rate of herbicids application (a.i kg/ha) or recommended dose}}{\text{a.i. in the formulation (\%)}} \times 100$$

Volume of water required (L/ha)

$$= \frac{\text{Volume of water used in test area (L)} \times 10000}{\text{Test area (width sprayed in m} \times \text{Length covered m)}}$$

Table 1. Classification of spray volume as per ISI norms

Class	Volume (L/ha)	Droplet size (micron)	Nature of spray	Equipments
High volume	> 560	> 400	Coarse	Hydraulic sprayer
Medium volume	56-560	201-400	Medium	Hydraulic sprayer
Low volume	5.6-56	101-200	Fine	Mist blower
Ultra low volume	0.56-5.6	100	Mist	Micron spryer or ULV Sprayer
Ultra-Ultra low volume	< 0.56	<50	Aerosol	Fogging machine, smoke and vapour generator

Examples

1. The calculate the quantity of Fluchloralin 40 EC if application rate is 1.0 kg. a.i. /ha.

$$=\frac{1.0}{40}\times 100 = 2.5 \text{ kg/ha}$$

2. Find the quantity of Printop 80 % W.P. (Simazine) required if applied @ 3.0 kg. a.i./ ha

$$=\frac{3.0\times 100}{80} = 3.75 \text{ kg/ha}$$

3. An amount of 10.2 liter of solution is required to spray of 150 sq.m. of wheat plot with 2,4-D. Calculate the volume of spray required for one ha.

$$=\frac{\text{Volume of water used in test area (L)}\times 10000}{\text{Test area (Width sprayed in m}\times\text{Length covered m)}}$$

$$=\frac{10.2\times 10000}{150} = 680 \text{ litre}$$

4. Calculate the amount of 2,4- D 40% required for 30 plots having an area of 20 sq.m. each. The recommended rate in the experiment is 2.0 liter a.i./ha.

Quantity of commercial formulations (herbicide) required (kg/ha)

$$=\frac{2.0}{40}\times 100 = 5.0 \text{ litre}$$

Quantity of commercial formulations required for test area in litre

$$=\frac{\text{Recommended dose in l/ha}\times\text{test area}}{10000} = 0.31 = 300 \text{ ml}$$

$$=\frac{5.0}{10000}\times 30\times 20 = 0.31 = 300 \text{ ml}$$

5. Calculate the volume of water required for one ha., if 150 m were covered while spraying a solution of 15 L with a swath width of 1.5 m , the volume required for one ha, would be:

$$=\frac{15\times10000}{150\times1.5}=\frac{150000}{225}=6661$$

6. Calculate the number of spray load if sprayer capacity is 15 liters. Also calculate the volume of water required for spraying 1 ha. using following observations.

A spray swath – 1.2 m.

Distance travelled – 60 m.

The volume sprayed in test area – 5 liters.

Amount of water required for 1 ha = $\frac{5\times10000}{1.2\times60}$ = 694 litres.

Number of sprayer loads/ha = $\frac{\text{Volume of spray solution required (l)}}{\text{Sprayer calacity (l)}}$

In the above example, if the sprayer capacity is 15 litres, then

Number of spray loads/ha = 694/15 =46

7. Calculate the concentration (%) of solution if 300 gram herbicide is mixed in 600 litre of water.

Concentration of solution (%)

$$=\frac{\text{Quantity of herbicides mixed (gms)}}{\text{Quantity of water (litre)}\times1000}\times100$$

$$=\frac{300}{600\times1000}\times100=\frac{3}{60}=0.05\%$$

8. Calculate the recommended dose of herbicide application (kg/ha), if commercial formulation of herbicides containing 50% a.i. was applied 3.0 kg per ha.

Quantity of commercial formulations (herbicide) required (kg/ha)

$$=\frac{\text{Rate of herbicides application (a.i. kg/ha) or Recommended dose}}{\text{a.i. in the formulation (\%)}}\times100$$

$$3.0=\frac{x}{50}\times100$$

$$x=\frac{3.0\times50}{100}1.5\text{ kg a.i./ha}$$

17

STUDY CROP WEED COMPETITION

INTRODUCTION

Weeds compete with crop plants for light, space, water and nutrients. The resultant effects are reduction in stand, vigour and yield (Table 1). Critical period is the duration of crop growth seriously infested with weeds and resultant into greater loss. Thus, elimination of weed crop competition during critical period of crop weed competition is necessary for optimal crop growth and maximum crop yield (Table 1). The basic objective to study the crop weed competition is to find out the number of monocot and dicot weeds associated with the particular crop, to estimate the competitive effect of weeds on crop growth, to study the weed and crop response to weed control treatments, to study efficacy of herbicides and to determine efficacy of intercropping on suppression of weeds.

Table 1. Critical period of crop weed competition and yield loss in different crops

Name of crop	Critical period of crop weed competition (Days)	Yield loss (%)
Cereals		
Rice (Direct sown)	15-45	15-90
Rice (Transplanted)	30-45	15-40
Wheat	30-45	20-40
Maize	15-45	40-60
Sorghum	15-45	15-40
Pearlmillet	30-45	15-60
Pulses		
Pigeonpea	15-60	20-40
Greengram	15-30	25-50

Name of crop	Critical period of crop weed competition (Days)	Yield loss (%)
Blackgram	15-30	30-50
Cowpea	15-30	15-30
Pea	30-45	20-30
Gram	30-60	15-25
Lentil	30-60	20-30
Oilseeds		
Soybean	20-45	40-60
Groundnut	40-60	40-50
Sunflower	30-45	30-60
Castor	30-50	30-45
Safflower	15-45	15-40
Sesame	15-45	15-40
Rapeseed and mustard	15-40	15-30
Linseed	20-45	30-40
Commercial crop		
Sugarcane	30-120	20-30
Potato	20-40	30-60
Cotton	15-60	40-50
Jute	30-45	
Vegetables		
Cabbage	30-45	50-60
Cauliflower	30-45	50-60
Okra	15-30	40-50
Tomato	30-45	40-70
Onion	30-75	60-70

Source: Tripathi, S.S.L., 2016

Need to measure weed density

It is necessary to know the number and nature of weeds present in field for competition with the crops. The weed density gives the biological stress that the crops are subjected to. It is easy to measure weed density during early stages (or) critical period of crop-weed competition.

Measurement of weed density

There are two methods normally employed for the measurement of weed density.

1. **Counting method**: Here different species of weeds are counted by using the quadrate in a crop field.
2. **Dry matter weight**: By this method, weeds cut at the base and their fresh and oven dry weight is recorded to compute the weed density.

Precautions

1. The weed count should be taken after the emergence of weeds but at the initial stage of crop growth to facilitate the placement of quadrate.
2. Take care to note the name and number of weeds immediately.
3. Select the location (or) site for placing quadrate at different places in crop field at random.

Materials required

- Quadrate of 0.5 m x 0.5m size (or any available size).
- Observation note book.
- Pencil.

Procedure

- Select a field with young crop and weed plants for weed count.
- Place the quadrate at randomly selected place in the field.
- First pull out all the dicotyledonous weeds inside the quadrate taking simultaneous counting.
- Record the number of dicotyledon weeds in the observation note book.
- Pull out and count all the monocotyledons and record them.
- Multiply the number of monocotyledonous/dicotyledonous weeds by 4 or 16 to convert it in unit area or m^2, if quadrate is having arm of 50 cm and 25 cm respectively.

Parameters: There are four major parameters to know the effectiveness/ efficiency of different weed control treatments:

A. Weed infestation =

$$\frac{\text{Total number of weeds} \times 100}{\text{Total number of weeds} + \text{Total number of crop plants}}$$

$$\text{Weed intensity} = \frac{\text{Total number of weeds of individual species} \times 100}{\text{Total number of weed plants}}$$

Problem- If a quadrate of 0.5 m^2 contains 24 weed plants and 32 of wheat plant. Calculate the weed infestation percentage in wheat

$$= \frac{24 \times 100}{24 + 32} = \frac{2400}{56} = 42.85\%$$

B. Weed index/Weed competition index: It refers to the reduction in crop yield in the presence of weeds in comparison to weed free crops. It is used to assess the efficacy of herbicide. Lower the value of weed index, better is the efficiency of herbicides and vice versa.

$$= \frac{\text{Yield in weed free plot (kg per ha)} - \text{Yield in weed infested plot (kg/ha)}}{\text{Yield in weed free plot (kg/ha)}} \times 100$$

Problem- If a quadrate of 50 m^2 of weed free plot of maize yielded 10 kg grain while it was 7.5 kg from weed infested plot. Calculate the weed competition index in maize

$$= \frac{\text{Yield in weed free plot} - \text{Yield in weedicide treated plot}}{\text{Yield in weed free plot}} \times 100$$

$$= \frac{10 - 75}{10} \times 100 = 25\%$$

C. Weed control efficiency: It indicates the efficiency of the applied herbicides or other practices to control weeds. It is the percentage reduction in weed dry matter under any weed control treatment in comparison to weedy check plot. This index is used to compare the different weed control treatments. Higher the WCE of any treatment, better is the treatment and vice versa. Either dry matter or count may be used for calculation (Annexure 14).

$$= \frac{\text{Weed population in control plot - Weed population in treated plot}}{\text{Weed population in control plot}} \times 100$$

Or

$$= \frac{\text{Weed dry matter in control plot (g)} - \text{Weed dry matter in treated plot (g)}}{\text{Weed dry matter in control plot (g)}} \times 100$$

Problem- A weed population in wheat plot of a quadrate of 80 m^2 was reduced to 120 as compared to 300 in adjoining plot of wheat remains untreated with Arelon. Find out the weed control efficiency of the herbicide.

$$= \frac{\text{Weed population in control plot (unweeded)} - \text{Weed pop. in treated plot}}{\text{Weed population in control plot}} \times 100$$

$$= \frac{300 - 120}{300} \times 100 = 60\%$$

Problem - A weed population in paddy plot of 150 m^2 was reduced to 42 as compared to 265 in adjoining plot of paddy remains untreated with Macheti (Butachlor). Find out the weed control efficiency of the herbicide.

$$= \frac{\text{Weed population in control plot} - \text{Weed population in treated plot}}{\text{Weed population in control plot}} \times 100$$

$$= \frac{265 - 42}{265} \times 100 = 84.15\%$$

D. Weed smothering efficiency: It is used to determine the effect of intercropping on suppression of weeds in comparison to pure stand crops

$$= \frac{\text{Dry matter of weeds from pure stand crop (g)} - \text{DM of weeds from intercropped plots (g)}}{\text{DM of weeds from pure stand crops (g)}} \; 100$$

Problem: In a field experiment with wheat the following observations on weed dry matter and yield were recorded by applying different herbicides. Calculate the weed control efficiency and weed index of each herbicide and recommend the best one to the farmer.

Treatment	Weed dry matter (g/m^2)	Crop yield (q/ha)	WCE %	Weed Index (%)
2,4-D	25	44	48-25/48 x 100=47.9	52-44/52 x 100=15.4
Metibuzin	28	38	48-28/48 x 100=41.6	52-38/52 x 100=26.9
Isoproturon	30	40	48-30/48 x 100=37.5	52-40/52 x 100=23.1
Fenoxaprop	25	35	48-25/48 x 100=47.9	52-35/52 x 100=32.7
Metsulfuron	20	50	48-20/48 x 100=**58.3**	52-50/52 x 100=**3.8**
Weedy check (No weed control)	48	17		
Weed free check	00	52		

Results

We will be recommended Metsulfuron to control weeds in wheat because it has lowest weed index value and highest weed control efficiency as compared to other herbicides.

18

STUDY OF HERBICIDE SPRAYERS

Introduction

The knowledge of sprayer and functions of their different parts is essential for proper maintenance of sprayer and effective application of herbicides. Herbicides are applied in the form of liquid sprays as well as solid granules depending on formulation.

Spray: It is defined as liquid discharged in particles and scattered as dispersed droplets.

Sprayer: It is an appliance which atomizes the spray fluid as spray. The spray fluid may be a suspension, an emulsion or solution.

Material required

Different types of sprayers having different capacity, Nozzles, Pressure gauge.

Kinds of sprayer

Knapsack sprayers: Knapsack sprayers are carried on the back of worker during spray operation. Usually they are metallic tanks, but now it is also available in plastic tanks. The plastic tanks are prone to damage by rats though light in weight. Knapsack sprayers may be hydraulic or pneumatic types of sprayers.

(i) **Hydraulic knapsack sprayers**: A hydraulic knapsack sprayer is manually operated sprayer which works under hydraulic pressure or energy to atomize and eject the spray fluid and offer constant pressure throughout the spray. Its tank capacity is usually of 15 litres, with provision for mechanical agitation of spray liquid. The worker uses his left hand to operate lever handle of the sprayer to maintain constant pressure. The hydraulic knapsack sprayer is primarily a high volume sprayer. The hydraulic knapsack sprayer could spray at 3 to 5 kg/ cm^2. Hydraulic knapsack sprayer is a low cost, easy to maintain and suitable for small holding. It is particularly satisfactory for spot treatment, band application of herbicides and blanket application. The main drawback of a hydraulic knapsack sprayer is its three way tiring action on man viz., a load on the back, one hand engaged in operating the pump, and other one in directing the lance.

(ii) **Pneumatic or compressed system knapsack sprayers:** Pneumatic knapsack sprayers are comparatively easy to work since they are pressurized before loading on the back of the worker. This allows a free hand to the operator. A big drawback of pneumatic sprayers is the decreasing spray work as spray advanced and may cause uneven spray.

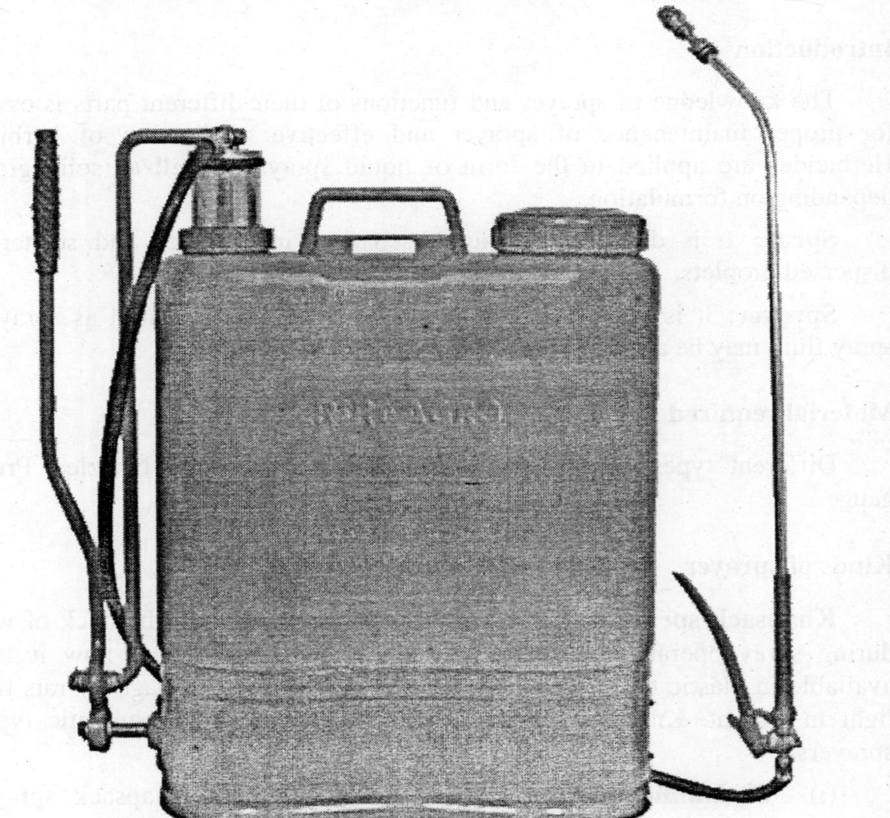

Fig. 1. Diagram of Knapsack Sprayer

(iii) **Motorized pneumatic sprayer -** It is a low volume sprayer suitable for spraying concentrated spray liquids. A blast of air acts as a carrier of herbicide concentrates in these sprayers which are called blowers. The air is forced through spraying jet of the delivery hose of blower and nozzle tube ejects the spray liquid. In this, blast in air atomizes the spray liquid into tiny drops. In this sprayer air acts as a carrier. The atomization is more vigorous when the air is pumped at faster rate in to the spraying unit. The herbicide is lost as spray drift as the blower produces the droplet size of 50-100 microns or mist blower. The main advantages of knapsack blowers are low volume spray resulting in reduction in loss of time in refilling the tank and fast spraying while low volume spraying with blowers is not as uniform as with high volume sprayers. The knapsacks are more preferably used for spraying of short translocated type of herbicides.

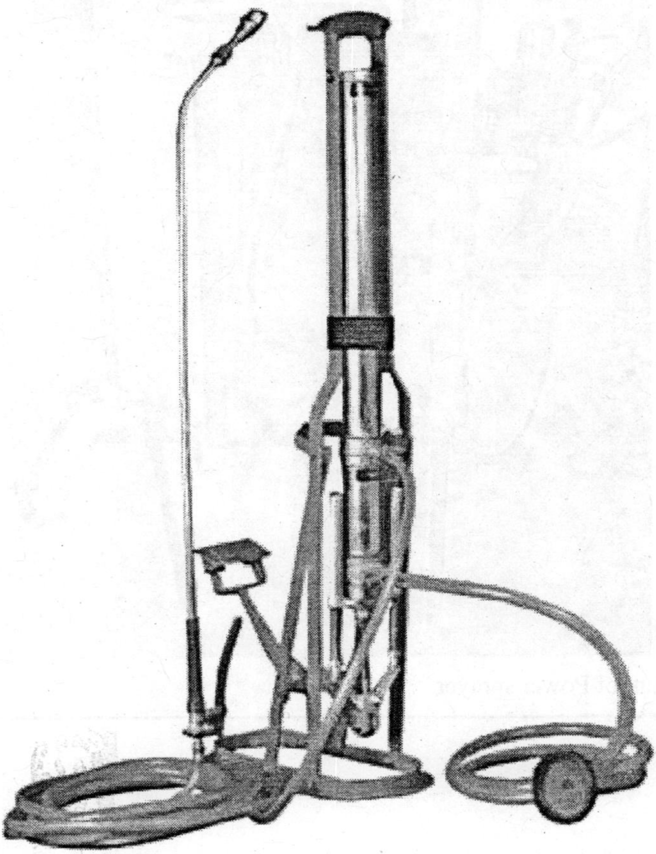

Fig. 2. Diagram of Foot Sprayer

Foot sprayers

The foot sprayer consists of a pump barrel and a pressure chamber. The plunger with a suction cup or piston drives into the pump barrel, thus sucking the liquid into the pressure chamber and expelling it through the discharge line. It is very popular for application of herbicides on comparatively large holding. Pump lever of a sprayer has a pedal. The main drawback of this sprayer is that it requires more than one person, normally three persons at a time.

Tractor powered sprayers

Tractor mounted sprayers work under a spray of 1.4-2.8 kg/cm^2. It is very useful equipment for large holding farmers. Tractor mounted sprayers give high uniformity of the spray and utilization of tractor during idle time.

Fig. 3. Diagram of Power sprayer

■■■■

19

SPRAY DISTRIBUTION SYSTEM OF SPRAYERS

INTRODUCTION

A distribution system of sprayer consists of nozzles, boom, spray lance (or spray boom) and hose. The main objective to study is to acquaint with different terms used in herbicide sprayers for efficient weed control.

(i) Nozzle: The function of spray nozzle is to break the pressurized spray liquid into fine droplets for application to the target weeds. Nozzles are identified by their droplet size, delivery or spray pattern that they produce.

There are at least six major types of nozzles used for spray of herbicides/ pesticides viz., (a) flat-fan nozzles, (b) solid cone nozzles, (c) flood-jet nozzles, (d) low volume nozzles, (e) centrifugal nozzles, and (f) blast nozzles. In general, smaller the spray droplet size, more the herbicidal phytotoxicity.

- **Flat-fan nozzle** - It lacks the vegetation penetrating capacity therefore, preferred for pre-emergence application of herbicides. However, it can also be used for post emergence application of the translocated type of herbicides which do not require penetration of the weedy vegetation.
- **Solid cone nozzle -** It provides good vegetation penetrating ability to the sprays, therefore, used in application of the contact type of herbicides on dense, weedy patches as spot treatment.
- **Flood-jet nozzle -** It produces medium droplet sprays and thereby reduce drift hazard. These nozzles are commonly used for sub-surface application of herbicides.
- **Low volume nozzle -** It is designed to apply herbicide concentrates without much dilution.
- **Centrifugal nozzle -** It meant for spraying viscous spray liquids like invert emulsions. Motorized mist blowers employ blast nozzles which feed the spray liquid into the air stream to split it into droplets and carry them as mist, off the delivery pipe.

- **Motorized (Blast) nozzle** - It feeds the spray liquid into the air stream to split it into droplets and carry them as mist off the delivery pipe.

Full cone (Hollow cone) nozzle

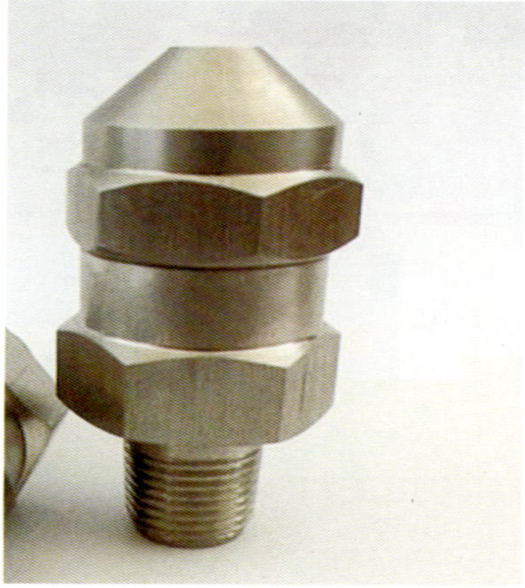

Full cone (Solid cone) nozzle

Flood jet nozzle

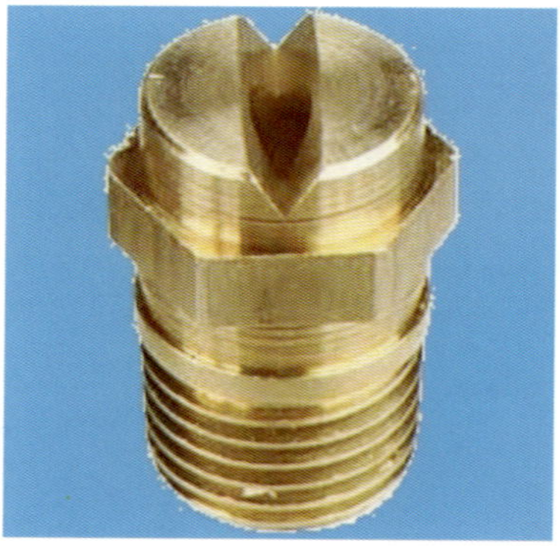

Flat fan nozzle

Blast nozzle

Mist nozzle

Fig. 1. Diagram of different types of nozzles

(ii) Lance: It is a brass tube, about 90 cm long, attached to delivery hose of the sprayer and fitted at its free-end with a replaceable nozzle.

(iii) Boom (spray bar): A boom is essentially a horizontal pipe with two to several nozzles on it. Usually, nozzles are spaced 50 cm apart on the bar.

(iv) Hose: It is made up of rubber and used to deliver spray fluid from the sprayers to lance or from container to the sprayer.

(v) Spray cut off device: To shut off flow of liquid into lance.

(vi) **Agitator:** It keeps the components of spray liquid in homogeneous mixture. It is essential for heterogeneous solutions like emulsions and suspensions.

(vi) Pressure regulators: Sprayers are fitted with pressure regulators with the objective to push the liquid at a constant desired pressure.

■■■■

20

NUMERICAL EXERCISES ON WATER REQUIREMENT

INTRODUCTION

Crop water requirement is defined as

- The quantity of water required by plants for their survival, growth and development to produce economic parts.
- It may be defined as the quantity of water required by a crop or diversified pattern of crop in a given period of time for its normal growth under field conditions at a place.
- Or is the quantity of water utilized by a crop, irrespective of its source for obtaining maximum yield in a particular area without any/minimum adverse effect on soil properties.

This requirement is supplied either naturally by precipitation or artificially by irrigation. Water dissolves the nutrients in soil and plant roots take up nutrient solution from soil and supply it to other parts of the plant. Nutrients and water support growth and development of the plants.

The estimations of the crop water requirement are essential for:

- Deciding the desired quantity of water required at right time during development of plants.
- Crop planning on the farm and
- Planning of any irrigation project to decide the stream size and canal capacity.

The crop water requirement includes:

a Transpiration loss through leaves (T)

b Evaporation loss through soil surface in cropped area (E)

c Amount of water used by plants (Wm) for its metabolic activities

d Other application losses (conveyance loss, percolation loss, runoff loss, etc.,) (WL).

e The water required for special purposes (WSP) like puddling, ploughing, land preparation, leaching of salts, for the purpose of weeding, for dissolving fertilizer and chemical, etc.

Hence the water requirement is symbolically represented as:

$$WR = T + E + Wm + WL + WSP$$

In a cropped field water evaporates (E) from the bare soil, is transpired through plants (T) and some quantity is retained in plants body for metabolic activities (Wm), added together (E+T+Wm) known as crop consumptive use of water (CU). Water retained in crop plants at any time is a very small fraction (less than 1% of the total water) to total quantity lost through evaporation and transpiration, together known as evapo-transpiration (ET). Both CU and ET, therefore, are used interchangeably.

$$Cu = E + T + Wm$$

Therefore,

$$WR = CU + WL + WSP$$

The other application losses and special purposes are mostly indented for wet land cultivation. Hence, under irrigated dry land crop, the ET losses alone are accounted for crop water requirement. Since, out of the total crop water requirement, some portion may be supplied by effective rainfall and contribution from soil moisture storage, irrigation water demand, therefore, equals the crop water demand *minus* the sum of effective rainfall and the moisture contribution from soil profile. It can also be stated based on "Demand" and "Supply source" as follows

$$WR = IR + ER + S$$

$$IR = WR - ER - S$$

Where,

IR - Irrigation requirement

ER - Effective rainfall

S - Contribution from ground water table.

If the crop water requirement of a particular crop is 6 mm per day, it means every day we have to give 6 mm of water to the crop. Practically it is not possible since it is time consuming and laborious. Hence it is necessary to schedule the water supply at sometime intervals. For example the water requirement of 6 mm/day can be scheduled as 24 mm/for every 4 days or 30 mm/for every 5 days or 36 mm/for every 6 days depending upon the soil type and climatic conditions prevailing in that particular place. While doing so we must be very cautious that the interval should not allow the crop to suffer for want of water.

The crop water requirement is estimated by following methods:

i. Transpiration ratio approach

ii. Depth-interval yield approach

iii. Lysimeters' studies

iv. Soil moisture depletion studies

v. Use of evaporimeters

vi. Use of empirical/energy balance method

The crop water requirement varies from place to place, crop to crop and depends on agro-ecological situations. Crop water requirement is governed by the following factors:

(1) Crop factors
  (a) Variety
  (b) Growth stages
  (c) Duration
  (d) Plant population
  (e) Crop growing season

(2) Soil factors
  (a) Structure
  (b) Texture
  (c) Depth
  (d) Topography
  (e) Soil chemical composition

(3) Climatic factors
  (a) Temperature
  (b) Sunshine hours
  (c) Relative humidity
  (d) Wind velocity
  (e) Rainfall

(4) Agronomic management factors
  (a) Irrigation methods used
  (b) Frequency of irrigation and its efficiency
  (c) Tillage and other cultural operations like weeding, mulching etc.
  (d) Intercropping etc.

Based on these factors, average crop water requirement for various crops have been worked out and given in Table 1.

Table 1: Water requirement of different crops

S.No.	Crop	Water requirement (mm)	S.No.	Crop	Water requirement (mm)
1	Rice	1200	15	Chick pea	250-300
2	Wheat	450-650	16	Mustard	300-350
3	Barley	300-350	17	Cumin/Isabgol	180-225
4	Maize	500-800	18	Peas/Onion	350-500
5	Sorghum	450-650	19	Cabbage	380-500
6	Pearl millet	250-300	20	Chillies	400-600
7	Khrif pulses	250-300	21	Potato	500-700
8	Groundnut	500-700	22	Tomato	600-800
9	Soybean	450-700	23	Ber/Aonla	600-650
10	Pigeon pea	450-500	24	Grapes	700-1200
11	Tobacco	400-600	25	Citrus	900-1200
12	Sugarbeet	550-750	26	Mango	1000-1200
13	Cotton	700-1300	27	Tumeric	1200-1400
14	Sugarcane	1500-2500	28	Bananas	1200-2200

Irrigation requirement

The irrigation requirement of crops refers to water requirement of crops exclusive of effective rainfall and contribution from soil profile. It depends on irrigation need of individual crop based on area of crop and losses in the farm water distribution system etc. It is usually expressed in terms of water depth per unit of land area (ha/cm) or unit of depth (cm). It may be expressed as:

$$IR = WR - (ER + S)$$

Where,

IR - Irrigation requirement

WR - Water requirement

ER - Effective rainfall

S - Soil moisture contribution

Net irrigation requirement

It is defined as the actual amount of irrigation water required to bring the soil moisture level in effective root zone to the field capacity, which in turn meets the ET demand of the crop. It is the difference between the field capacity and soil moisture content in the root zone before starting irrigation. To work out the net irrigation requirement, ground water contribution and other gains in soil moisture are to be excluded.

$$d = \sum_{i=I}^{n} \frac{M\,fci - M\,bi}{100} \times Ai \times Di$$

d = Net irrigation water to be applied (cm)

Mfci = Moisture content at FC in ith layer (%)

Mbi = Moisture content before irrigation in ith layer (%)

Ai = Bulk density (g/cc)

Di = depth (cm)

n = number of soil layer

Gross irrigation requirement

The total quantity of water used for irrigation is termed gross irrigation requirement. It includes net irrigation requirement and losses during water application and other losses. The gross irrigation requirement can be calculated for a field, for a farm, for an outlet command area, and for an irrigation project. The gross irrigation requirement depends on the need after considering the approximate losses at various stages of crop.

$$\text{Gross irrigation} = \frac{\text{Net irrigation requirement}}{\text{Field efficiency of system}}$$

Irrigation frequency/Irrigation interval

Irrigation frequency is an interval between two consecutive irrigations without rainfall during highest consumptive use, i.e. peak consumptive use period of crop and expressed in days. It depends upon the rate of uptake of water by crop plants, soil moisture supplying capacity and soil moisture available in the root zone. Hence, it is a function of crop, soil and climate. Normally, irrigation should be given at about 50 per cent and not over 60 per cent depletion of the available moisture from the effective root zone.

$$\text{Irrigation frequency (days)} = \frac{\text{Field capacity} - \text{Moisture content in root zone prior to starting irrigation}}{\text{Peak period consumptive use rate of the crop}}$$

$$= \frac{\text{Net depth of application}}{\text{Peak period moisture use rate}}$$

Irrigation period

Irrigation period is the number of days that can be allowed for applying one irrigation to a given design area during peak consumptive use period of the crop.

$$\text{Irrigation period (days)} = \frac{\text{Net amount of moisture in soil prior to starting irrigation} - \text{lower limit of depletion (PWP)}}{\text{Peak period consumptive use rate of the crop}}$$

Example 1: Find out the net quantity of irrigation water to be applied to wheat field with the following values of moisture status.

Depth of soil layer in (cm)	Moisture% on oven-dry basis		Bulk density g/cc
	Field capacity	Actual	
0-15	25.0	16.4	1.39
15-30	24.0	17.8	1.47
30-60	22.3	19.2	1.51
60-90	22.2	20.5	1.53

Solution

Moisture deficits in the different layers will be as follows:

$$1^{st} \text{ layer} = \frac{25.0-16.4}{100} \times 1.39 \times 15 = 1.79 \text{ cm}$$

$$2^{nd} \text{ layer} = \frac{24.0-17.8}{100} \times 1.47 \times 15 = 1.36 \text{ cm}$$

$$3^{rd} \text{ layer} = \frac{22.3-19.2}{100} \times 1.51 \times 30 = 1.40 \text{ cm}$$

$$4^{th} \text{ layer} = \frac{22.2-20.5}{100} \times 1.53 \times 30 = 0.78 \text{ cm}$$

Total deficits = 5.33 cm

Net quantity of water to be applied is 5.33 cm.

21

CRITICAL STAGES APPROACH FOR IRRIGATION SCHEDULING

INTRODUCTION

Critical stage is the stage at which the water stress causes a considerable reduction in yield. It is also known as moisture sensitive period. Moisture stress due to restricted supply of water during critical stage will irrevocably reduce the yield even provided with adequate water and fertilizer at other growth stage. At critical stages, optimum water level should be maintained through timely irrigations. In general for most of the agronomic crops, the mid season stage is most sensitive to water shortage while the least sensitive stages are ripening and harvesting. Under scarce condition, in an irrigation project or in a farm, if mono cropping is followed with staggered sowing or planting, it is better to schedule irrigation such that crop coincides with most critical stage. The sensitive stages vary from crop to crop as given below.

Table 1: Sensitive stage of different crops

Crop	Critical stages / Sensitive stages
Cereals and millets	
Rice	Panicle initiation, heading and flowering
Sorghum	Flowering and grain formation
Pearlmillet	Heading
Maize	Tasseling, silking and grain filling
Wheat	Crown root initiation, tillering , booting milk and dough
Barley	30 DAS (Tillering)
Pulses	
Chickpea	Pre-flowering and pod development
Pigeon pea	Flowering initiation, pod development

Crop	Critical stages / Sensitive stages
Greengram	Flowering initiation, pod development
Cluster bean	Branching and flowering
Oil seeds	
Groundnut	Flowering peg initiation and penetration and pod development
Sesame	Blooming to maturity
Mustard	Pre-flowering and pod development
Sunflower	Two weeks before and after flowering
Soybean	Blooming and seed formation
Safflower	From rosette to flowering
Castor	Full growing period
Cash crops	
Cotton	Flowering and Boll formation
Sugarcane	Maximum vegetative stage
Tobacco	Immediately after transplanting
Potato	Tuber initiation and maturity
Forage	
Alfalfa	Immediately after cutting for hay crop and flowering for seed crop

■■■■

22

IRRIGATION SCHEDULING BASED ON IW/CPE APPROACH

INTRODUCTION

Irrigation scheduling is a systematic method of deciding as to when and how much water to apply to replenish the soil moisture at the desired level for maximizing crop yields and optimize water use efficiency. Criteria most suitable for scheduling irrigations vary from one situation to another. Planning optimum irrigation schedules requires information or knowledge on following set of conditions as to when and how much water stress occurs:

i. soil hydraulic properties, depth, salinity and nutrient status of root zone,

ii. ground water level,

iii. method of water application,

iv. availability and quality of water,

v. soil water status,

vi. weather conditions,

vii. plant characteristics and status of crop stress,

viii. potential yield reduction if the crop remains in a stressed condition and

ix. other specific needs such as crop cooling, frost protection etc. must be considered.

Advantages of irrigation scheduling

- enables the farmer to rotate the irrigation schedule among various fields to minimize crop water stress and maximize yield.
- helps to reduce water losses and minimize waterlogging problems.
- enhances the efficiency of other inputs viz., fertilizers, pesticides, herbicides etc.

- help to reduce energy and labour requirement by avoiding over-irrigation.
- helps in maintaining soil health.
- water saved can be used to irrigate additional area.
- increased net profits by increasing crop yields and quality.

Approaches of irrigation scheduling

Irrigation scheduling is done based on many criteria to learn better management of irrigation water for obtaining maximum production. It requires monitoring various indicators to determine the need for irrigation.

- Soil water status (soil moisture measurement)
- Weather parameters (climatological indices)
- Plant parameters (plant indicators)
- Simulation modelling approach

The irrigation scheduling programme provides the possibility to:

- Develop and plan indicative irrigation schedules adapted to field operational conditions;
- Evaluate field irrigation programmes in terms of efficiency of water use and yield production;
- Simulate field irrigation programme under water deficiency conditions, rainfed conditions, supplementary irrigation, etc.

It is well known that the consumptive use of water is physical phenomena governed by the incident energy at a place. Considering the strong relationship between the pan evaporation and crop ET, particularly in semi-arid tropics, the IW/CPE ratio has been suggested by Prihar et al. (1984). It is a scientific and widely adopted climatologically approach for irrigation scheduling. In this, IW/CPE ratio, i.e., the ratio of a fixed amount of irrigation water (IW) and cumulative pan evaporation exclude effective rains (CPE), since previous irrigation is used for scheduling irrigation to crops. Thus, this approach considers both available water holding capacity of soils and climatic factors for scheduling irrigation and provide knowledge on correct time (when) and optimum quantity of water (how much) to be applied at each irrigation. The depth of irrigation water (IW) is decided by the available water holding capacity of the soil and the probable deficit before irrigation and thus desired irrigation efficiency can be achieved. It is possible to make recommendations on irrigation schedule for a crop by using the information on percentage of permissible depletion of available water for the crop, field capacity and permanent wilting point values of the soil, the daily mean pan evaporation of the location, and the crop factor or IW/CPE ratio. Soil and rooting depth of the crop also would be required additionally. Thus it optimizes crop yield with maximum water use efficiency and minimum damage to soil properties besides efficient use of water and energy. The ratio IW/CPE serves as a soil moisture stress index.

Material required

USWB Class A pan evaporimeter, measuring cylinder, rain gauge and water source.

Procedure

- Find out the permissible soil water deficit without affecting the yield.
- Find out the irrigation depth on the basis of permissible soil water depletion from the root zone for potential yield. For example, if the plant available water (PAW) value is 8 cm and the allowable depletion is 50 per cent, depth of irrigation shall be 4 cm.
- Find a factor (known as IW/CPE ratio) which is to be multiplied with CPE to compute the depth of water depleted from the soil storage.
- Irrigation is applied as soon as CPE occurs to a predetermined level or multiplied with IW/CPE ratio becomes equal to the irrigation depth. Days required to cumulate this much CPE will differ according to the climatic conditions.

Limitations

- This criterion can easily be adopted by Indian farmers but the lacunae is as it varies from region to region and soil to soil.
- This is used to fix the net irrigation (IW) in cm. Gross depth of irrigation shall be fixed taking into consideration irrigation efficiencies.
- This criterion does not consider the critical growth stages for irrigation.
- This method may also not be well for crops requiring 1-2 irrigations only, in timing of first and last irrigation which are under low temperature/frost/salt stresses.

For wheat, IW/CPE ratio is from 0.75 to 1.00. For example, an IW/CPE ratio 1.0 with 4 cm irrigation means that whenever cumulative pan evaporation values reach 40 mm, irrigation water of 4 cm is applied. Note that IW/CPE ratio and crop-coefficients (Eta /ETo) are not equal values.

Table 1. The optimum ratio of IW/CPE for major crops.

Crop	Crop factor
Pulses/oil seeds	0.9-1.0
Rice	1.2-1.4
Wheat	0.75-1.00

The following are some worked out examples for arriving at irrigation schedules.

Example 1.

In a place, the IW/CPE ratio of wheat is found to be 0.75. If the depth of irrigation is 50 mm, calculate the cumulative pan evaporation values required for starting irrigation.

Solution:

Given: IW/CPE=0.75 and depth of irrigation = 50 mm

Thus: 50/CPE = 0.75

CPE= 50/0.75= 66.6 mm.

Conclusion

Irrigation at 50 mm is given when the sum of daily pan evaporation values reaches to 66.6 mm. In the table given below, CPE reaches 66.6 mm on 8th December. Therefore, irrigation will be done on 8th December.

Table 2. Pan evaporation data

Month	Date	Pan evaporation/ day (mm)	Cumulative pan evaporation (mm)
November	15	2.5	2.5
	16	2.4	4.9
	17	2.6	7.5
	18	2.5	10.0
	19	2.6	12.6
	20	1.3	13.9
	21	1.9	15.8
	22	1.2	17.0
	23	4.5	21.5
	24	3.6	25.1
	26	1.9	29.2
	27	3.4	32.6
	28	2.4	35.0
	29	2.2	37.2

Month	Date	Pan evaporation/ day (mm)	Cumulative pan evaporation (mm)
	30	2.7	39.9
December	1	1.6	41.5
	2	2.6	44.1
	3	3.9	48.0
	4	5.5	53.5
	5	4.9	58.4
	6	4.6	63.0
	7	2.2	65.2
	8	1.5	**66.7**
	9	1.8	68.5
	10	2.4	70.9
	11	5.1	76.0

Example 2

Calculate the amount of water for each irrigation for scheduling irrigation at IW/CPE – 0.5 with 10 cm of CPE

IW/CPE = 0.5

IW/10 = 0.5

IW = 0.5x10= 5cm

Example 3

Calculate the irrigation internal for a 70 day old maize crop using following data.

Rooting depth – 60 cm

Soil type – loamy sand

FC – 16 %

PWP – 6%

Allowable soil moisture depletion – 50%

Irrigation application efficiency – 75%

Peak daily evaporation – 6 mm

Kc value – 1.06

PAW = 16 – 6 =10%

PAW per 60 cm = 10/100 × 60 = 6 cm

Allowable depletion = 6 × 50/100 = 3 cm

∴ Net depth = 3 cm

Gross depth = 300/75 = 4 cm

Daily water use by the crop = 6 × 1.06 = 6.36 mm

Interval of irrigation = 30/6.36 = 4.72 ≈ 5 days

$$= \frac{\text{Allowable depth}}{\text{Daily water use by the crop}} = \frac{30\text{ mm}}{6.36\text{ mm}} = 4.72 = 5\text{ days}$$

■■■■

23

SCHEDULING OF IRRIGATION BASED ON SOIL MOISTURE DEPLETION APPROACH

INTRODUCTION

To achieve better productivity, it is important to work out an efficient and economic irrigation schedule for water use under any given set of agro-climatic conditions. There are several approaches for scheduling irrigation based on crops, soil, atmosphere and plant water relations as discussed in previous experiment. According to soil moisture depletion concept, the water content at FC is considered as 100 per cent available for crop growth and that at PWP is 0 per cent. The safe limit of allowable soil water depletion for a crop is determined by experimentation and taken as criterion for scheduling irrigation. Values for FC and B.D. are determined layer wise of a soil profile to workout the moisture deficit with the formula:

$$d = \frac{\text{Field capacity} - \text{Permanent wilting point}}{100} \times \text{Bulk density} \times \text{Depth of soil}$$

Based on this principle a large number of irrigation experiments have been conducted and it was observed that most of the crops like wheat, barley, maize, bajra and sunflower can resist the moisture stress upto 50% of available water in the root zone (Annexure 21).

Example 1

In an irrigation trial on wheat, irrigation is to be scheduled at 20, 40, 60 and 80 per cent available soil water in 0-30 cm depth. If the FC is 28 per cent and PWP is 12 per cent, calculate the moisture present at which irrigations would be scheduled under different irrigation treatments.

Solution

I_1 = Irrigation at 20% ASW = (28-12) x 20 / 100 + 12 = 15.2%

I_2 = Irrigation at 40% ASW = (28-12) x 40 / 100 + 12 = 18.4%

I_3 = Irrigation at 60% ASW = (28-12) x 60 / 100 + 12 = 21.6%

I_4 = Irrigation at 80% ASW = (28-12) x 80 / 100 + 12 = 24.8%

Example 2

Taking Bikaner data, the example to calculate irrigation interval is given below:

FC	20.0% dry weight basis
PWP	8.0% dry weight basis
Bulk density	1.40 g/cc
Root depth	60 cm
ET rate	0.5 cm/day

Allowable soil water depletion is equal to 25 per cent of available soil water.

Solution

$$d = \frac{20.0-8.0}{100} \times 1.40 \times 60 = 10.08 \text{ cm}$$

$$\text{Allowable soil-water depletion} = \frac{25 \times 10.08}{100} = 2.52 \text{ cm}$$

$$\text{Interval of irrigation} = \frac{2.52}{0.5} = 5.04 = 5 \text{ days}$$

In the present situation irrigation should be given at 5 days interval.

■■■■

24

IRRIGATION EFFICIENCY

INTRODUCTION

The term 'efficiency' refers to output per unit of input. Evaluation of performance in terms of efficiency is prerequisite for proper use of irrigation water. There are considerable losses of irrigation water during its application and distribution in the field. In most irrigation projects, the irrigation efficiency ranges between 12 and 34%. An efficient irrigation system implies effective transfer of water from the source to the field with minimum possible loss. The objective of study the efficiency concept is to identify the nature of water losses and types of action taken for improvements in the system. Keeping these losses in view, the actual amount is greater than the net amount of water needed in the soil. Irrigation efficiency or overall efficiency indicates how efficiently or beneficially the available water supply is being used. It refers to the ratio usually expressed as percent of the total volume of irrigation water actually used by the plants in transpiration, evaporation from soil, for regulating salt concentration and metabolic activities of the plants to the total volume of water diverted, stored or pumped for irrigation from a point of measurement or simply the ratio of how much water utilized to how much water applied. It can be expressed as

$$E_i = \frac{W_{et} + W_{salt} - ER}{Wi} \times 100$$

Where

E_i= Irrigation efficiency

W_{et} = Volume of irrigation water per unit area of land transpired by the plants, evaporated from the soil during the crop period (including field preparation and nursery).

W_{salt}= Volume of irrigation water per unit area of land to regulate salt content of soil

ER = Effective rainfall

W_i = Volume of water per unit area of land that is stored in a reservoir for irrigation purpose or diverted for irrigation.

Types of irrigation efficiencies

For calculation purpose, the term may include either the entire irrigation project or any component of it depending upon the physical, biological and economic importance/ comparisons. As the irrigation water is lost at several stages during the irrigation, the irrigation efficiency can be monitored at various stages of the irrigation process. The different types of irrigation efficiencies commonly used are based on analysis of numerator and denominator points and can be classified into following categories:

(i) Water conveyance efficiency

(ii) Water application efficiency

(iii) Water storage efficiency

(iv) Water distribution efficiency

(v) Project irrigation efficiency

(vi) Operational irrigation efficiency

(vii) Economic irrigation efficiency

Water conveyance efficiency

Water losses which occur in conveyance from the point of diversion to till it reaches the farmer's fields can be evaluated by water conveyance efficiency. These losses are due to seepage and evaporation in the canal network, water course and field channels. Weeds and aquatic plants along and in the irrigation channels also significantly decrease the amount of water. Thus, water conveyance efficiency is the ratio of water delivered in fields at the outlet head (water courses and field channels) to that diverted into the canal system from the river or reservoir. It is related to the conveyance sources such as canal networks, water courses, farm channels and of water conveyance from an individual well/ tubewell to a field.

$$E_c = \frac{W_t}{Wf} \times 100$$

where,

Ec = water conveyance efficiency (%),

W_f= water delivered to the farm by conveyance system (at field supply channel)

W_t = water introduced into the conveyance system from the point of diversion

Water conveyance efficiency is generally low as about 21% losses occur in earthen water courses only. The losses depends on the nature and porosity of soil, depth, turbidity and temperature of water, age and shape of the canal section and ground water level etc. The losses can be minimized by lining the conveyance system and cleaning weeds and aquatic plants.

Water application efficiency

It is the ratio of volume of water that is stored in the root zone of crops (ultimately consumed by transpiration or evaporation or both) to the volume of water actually delivered at the field head. It is a measure of efficiency of water application in the field. In general, water application efficiency decreases as the amount of water increases in progressive irrigation. It is also termed as farm efficiency as it takes into account water lost in application at the farm. Water losses due to inefficient application of water in the field vary from 28 to 50%.

$$E_a = \frac{W_s}{W_f} \times 100$$

where,

Ea = water application efficiency, per cent

Ws = irrigation water stored in the root zone of crops

Wf = irrigation water delivered to the farm (at field supply channel)

Common sources of loss of irrigation water during application are surface runoff from the farm and deep percolation in soil below the root zone.

Water storage efficiency

It is considered that irrigation with small quantities of water may lead to high water application efficiencies. The water storage efficiency confirms this concept and is defined as the ratio of the water stored in the root zone by irrigation to the water needed in the root zone to bring it to the field capacity (potential soil moisture storage volume). It refers to how completely the water needed prior to irrigation has been stored in the root zone during irrigation thus its objective is to refill the root zone. It is also termed as water storage factor or water requirement efficiency (ratio of the amount of water beneficially used during an irrigation interval to the requirement during that interval).

$$E_s = \frac{W_s}{W_n} \times 100$$

Where,

Es = water storage efficiency, per cent

W_s = water stored in the root zone during the irrigation

W_n = water needed in the root zone prior to irrigation.

Water storage efficiency assumes importance when water supplies are limited or when excessive time is required to secure adequate penetration of water into the soil. This parameter is the most directly related with the crop yield since it reflects the degree of soil moisture stress. When salt problem is there, water storage should be kept high to maintain a favorable salt balance. The value of the storage efficiency is important when either the irrigation tends to leave major portions of the field under-irrigated or where under irrigation is purposely practised to use the precipitation as and when it occurs.

Water distribution efficiency

Uniform distribution of water over the entire field is necessary for high crop production. Water distribution efficiencies indicate the extent to which water is uniformly distributed along the run. The length of run in furrows or borders is decided to a large extent by the uniformity of water distribution for a given soil, size of stream and management practice. Expression for distribution efficiency to evaluate the extent to which the water is uniformly distributed is as follows:

$$\text{Ed} = \left(1 - \frac{y}{D}\right) \times 100$$

where,

Ed = water distribution efficiency, per cent

d = average numerical deviation in depth of water stored from average depth stored during irrigation

D = average depth of water stored along the run during irrigation

A water distribution efficiency of 80% means that 10% of water was applied in excess and consequently 10% was deficient in comparison to the average depth of application.

Project irrigation efficiency

Project irrigation efficiency (Ep) is the ratio of average depth of water stored in the root zone (available for consumptive use by the crop) during irrigation (Ws) to the water delivered from the reservoir (Wr).

$$E_p = \frac{W_s}{W_r} \times 100 = \frac{E_a \times E_c}{100}$$

Economic Irrigation efficiency

It is the ratio of the total farm production (net or gross profit) attained with operation of irrigation system, compared to the total production expected under ideally operated system under ideal conditions. Since this relates to final input and output it is a measure of the overall efficiency of the project.

Consumptive use efficiency

It is defined as the ratio of consumptive water use by the crop of irrigated farm or project and the net amount depleted from root zone of the soil on the farm or the project area. After irrigation water is stored in the soil, it may not be available for use by the crop because water may evaporate from the ground surface or continuously move downward beyond the root zone as it may happen in wide furrow spacing. The loss of water by deep penetration and by surface evaporation following irrigation is evaluated from the following expression:

$$E_{cu} = \frac{W_{cu}}{W_d} \times 100$$

Where,

Ecu = consumptive use efficiency, per cent

W_{cu} = normal consumptive use of water

W_d = net amount of water depleted from root zone soil

Consumptive use efficiency is useful in explaining the difference in crop response from different methods of irrigation.

Deep percolation ratio

The loss of water through drainage beyond the root zone is reflected in the deep percolation ratio (DPR). It is measured as under:

$$\text{DPR} = \frac{\text{Volume of deep percolation}}{\text{Volume of water applied to the field}} \times 100$$

High deep percolation losses aggravate water logging and salinity problems and leach valuable crop nutrients from the crop root zone. These losses can return to receiving streams heavily laden with salts and other toxic elements and thereby degrade the quality of water to be used by others or eutrophication.

Tail water ratio

Losses of irrigation water from the irrigation system through surface runoff from the end of the field are indicated in the tail water ratio (TWR).

$$\text{TWR} = \frac{\text{Volume of surface runoff}}{\text{Volume of water applied to the field}} \times 100$$

Runoff losses pose additional threats to irrigation systems and regional water resources. Erosion of the top soil on a field is generally associated with the run-off. The sediments can then obstruct conveyance and control structures downstream, including dams and regulating structures.

Operation efficiency

It is the ratio of actual project efficiency compared to the operational efficiency of an ideally designated and managed system using the same irrigation method and facilities. Low operational efficiencies indicate defective management of system design.

Water use efficiency

Water resource of India is limited and hence available water is to be used in the most efficient manner possible. Having conveyed water to the point of use and having applied it, the next efficiency concept of concern is the efficiency of water use. Water use efficiency (WUE) is the yield of marketable crop produced per unit of water used and is expressed as (kg/ha-cm).

Water use efficiency is also defined as (i) crop water use efficiency and (ii) field water efficiency.

(a) Crop Water Use Efficiency: It is the ratio of yield of crop (Y) to the amount of water depleted by crop in evapotranspiration (ET). It is expressed in kg/ha/mm or kg/ha/cm.

$$E_{cwu} = \frac{Y}{ET} = \frac{Y}{G+E+T} = \frac{Y}{CU} = \frac{Y}{ET+D}$$

Where,

E_{cwu} = Crop water use efficiency

Y = Crop yield

ET = Evapotranspiration

G=Amount of water taken up and used for crop growth,

E= Evaporated directly from the soil surface,

T= Transpired through foliage

Cu = Consumptive use

(b) Field Water Use Efficiency:

It is the ratio of yield of crop (Y) to the total amount of water used in the field. It is expressed in kg/ha/mm (or) kg/ha/cm

$$E_{fwu} = \frac{Y}{WR} = \frac{Y}{G+E+T+D}$$

Where,

E_{fwu} = Field water use efficiency

WR = Water requirement

G = Amount of water taken up and used for crop growth,

E = Evaporated directly from the soil surface,

T = Transpired through foliage

Cu = Consumptive use

D = Deep percolation losses.

Deep percolation is important for rice while seepage is important for other crops. Of the two indices defined, the crop water use efficiency is more of research value whereas the field water use efficiency has greater practical importance for planners and farmers.

Significance of irrigation efficiencies

In general, a low value of any of the irrigation efficiency implies that the land, water and crops are not being managed properly. A low value of conveyance efficiency expresses that much of the water released from the source is lost in transit from source to the field. Low application efficiency means wastage of water in the form of deep percolation or runoff losses. Poor storage efficiency means that water has been applied inadequately. A poor distribution efficiency results due to uneven land surface. There are low patches where water will penetrate more and there are high patches where water can not reach. A low water use efficiency also results due to over application of water or inability of the crops to utilize the applied water due to poor vegetative growth or adverse chemical properties of root zone soil and water. The net effects of poor irrigation efficiencies are crop loss and wastage of water and nutrients.

Factors Affecting Irrigation Efficiency

i. Soil texture: Finer the soil texture, higher the surface run-off and lesser the deep percolation losses compared with coarse sandy soils.

ii. Land topography: Irregular land surface, compact shallow soils reduce the irrigation efficiency

iii. Stream size: irrigation efficiency will be low with small or too large irrigation streams

iv. Frequency of irrigation: Long irrigation runs and excessive single applications contribute to large losses of irrigation water.

Example 1

Calculate the farm conveyance efficiency and field water application efficiency when a stream of 90 L/s received at the farm gate after being diverted from a canal delivered 68 L/to the field. During irrigation to wheat crop for 8 h, 320 and 150 m^2 of water respectively were lost by runoff and deep percolation.

$$E_c = \frac{W_t}{W_f} \times 100 = \frac{68}{90} \times 100 = 75.5\%$$

Water lost during irrigation (L) = runoff +deep percolation

= 320 + 150=470 m^3= 470 × 1000= 470000

Water delivered to the field = 68 L/s for 8 hours

= 68 × 8 × 60 × 60 = 1958400 L

Water stored in the root zone = water delivered -- losses

= 1958400 – 470000= 1488400 L

Therefore,

Water application efficiency =Water stored in the root zone/ water delivered to the field × 100

= 1488400/1958400 × 100= 76%

Example 2

If 10% water is lost in irrigation channel, 20% during application and 25% during distribution, it means the water conveyance efficiency, water application efficiency and water distribution efficiency are 90%, 80% and 75%, respectively.

The irrigation efficiency in this case will be

$$= Ic \times Ia \times Id = 0.90 \times 0.75 \times 0.80 = 0.54 \text{ or } 54\%.$$

■■■■

25

MEASUREMENT OF IRRIGATION WATER AT FARM

Introduction

Water is today's very precious and valuable input in irrigated agriculture. The planning of cropping systems and design of irrigation method depends on the available discharge from irrigation projects. Systematic and accurate water measurement and its interpretation help in application of water in accordance to crop water requirement for enhancing irrigation efficiency resultant into higher yields. Water is conveyed through open channels or pipelines from different sources to farms. Due to lack of knowledge about suitable water measurement, there are possibilities of either over-irrigating or under-irrigating the crops results in wastage of valuable resource, chances of crop damage and uneconomic returns. The measurement of discharge is also important in successful operation of a distribution system. Further, it is also important if the government agencies desire to charge the water rates to the farmers on volume basis. All the devices give sufficiently accurate results when they are constructed, installed and operated properly. The choice between one and other depends on the expected flow rates and site condition.

Water can be measured on volume basis if it is at rest or on discharge basis if it is in motion. Different units are used to measure irrigation water at rest and while it is flowing. Large quantity of water present in a reservoir may be measured in m^3. Volume of water applied to a field can be measured as ha-cm.

Unit of Measurement

1. At rest: Measured in units of volume such as litre, cubic metre, hectare centimetre (ha cm) and hectare metre (ha-m) (Annexure 11)

- **Litre:** A volume equal to one cubic decimetre or 1/1000 cubic metre.
- **Cubic metre:** A volume equal to that of a cube i.e 1 metre long, 1 metre broad and 1 metre high (1 cubic metre = 1000 litres).

- **Hectare-centimetre:** A volume necessary to cover an area of 1 hectare or 10,000 sq m to a depth of 1 centimetre (1 hectare centimetre= 100 cubic metres = 100, 000 litres).
- **Hectare-metre:** A volume necessary to cover an area of 1 hectare to a depth of 1 metre (1 ha-m = 10,000 cubic metres = 10 million litres).

2. Flowing water: It is the volume of water passing through a section per unit time. The units of measurement are such as litres per second, litres per hour, cubic metres per second, hectare centimetres per hour and hectare metres per day.

- **Litre per second:** A continuous flow amounting to 1 litre passing through a point in each second (generally used to denote the discharge of a pump, small stream or pipeline).
- **Cubic metre per second:** A flow of water equivalent to a stream 1 metre wide and 1 metre deep flowing at a velocity of 1 metre per second.

The conversion factors for measurement of water at rest and flowing conditions are given in Annexure 11.

Water measurement methods:

Various methods are used to measure the water and can be grouped into following categories:

(i) Volumetric methods

(ii) Velocity area method

(iii) Direct measurement structures like weir, orifice, parshall flame etc. and

(iv) Tracer techniques

(i) Volumetric method

It is the most simple and accurate method used to measure small discharge flowing through a channel or a pipe. In this method, irrigation stream (water) is collected in a container of known volume for a measured period of time with the help of a stop watch. In case of flow through a channel, some drop is necessary to allow the container to be placed below the channel level for collection of the discharge. The discharge rate is then calculated by noting the time for the reservoir to fill to a certain depth or for the water surface to rise to a predetermined level and rate of flow is calculated by the formula as (Fig. 1).

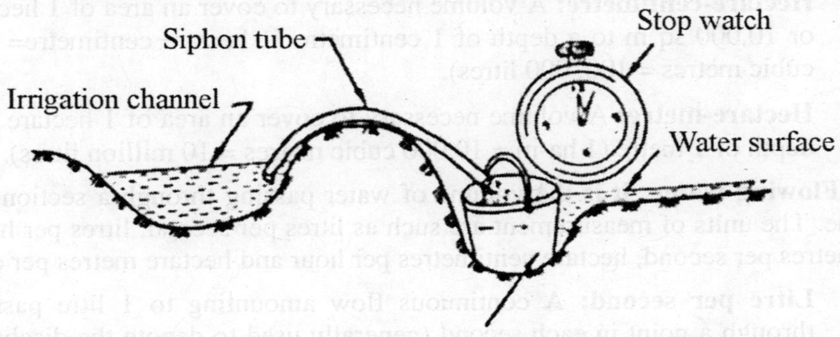

Fig. 1. Volumetric method of water measurement (Courtesy: Rao, V.P. *et al.*, 2010)

$$\text{Discharge rate}\left(\frac{\text{litre}}{\text{sec. or min.}}\right)=\frac{\text{Volume of water in container (liter)}}{\text{Average time taken to fill (sec. or min.)}}$$

Example 1.

Calculate the discharge, if a barrel of 210 litres capacity is filled in 30 seconds

Solution:

$$\text{Discharge rate}\left(\frac{\text{litre}}{\text{second}}\right)=\frac{\text{Volume of water in container (l)}}{\text{Average time taken to fill (sec.)}}=\frac{210}{30}=7.0$$

Example 2

A 20 litre capacity bucket is filled in 10 seconds by the discharge from a persian wheel, then the rate of flow will be:

$$\text{Discharge rate}\left(\frac{\text{litre}}{\text{second}}\right)=\frac{\text{Volume of water in container (liters)}}{\text{Average time taken to fill (sec.)}}=\frac{20}{10}=2.0$$

(ii) Velocity – area methods

It involves the measurement of velocity of water flow passing a point in an open channel. The flow rate is estimated by multiplying the cross-sectional area of the flow right angles to the direction of flow by the average velocity of water.

Discharge = Area x Velocity

$Q = a \times v$

Where,

Q = discharge rate, m^3/sec, cm^3/sec, l/sec

a = Area of cross section of channel or pipe, m^2, cm^2

v = velocity of flow, m/sec, cm /sec.

The cross sectional area is determined by direct measurements while velocity is generally measured by float method or current meter (Annexure 15).

1. Float method

This is one of the simplest methods and gives appropriate velocity of flow in a channel. Any object that flows in water but does not obstruct wind or gets carried away by it, can be used as a float. A straight length of a stream about 30 m long such that the cross-section is almost uniform through the length is selected and a block of about 1 cm^3 size of drywood serves as a float. The time taken by float to travel through a known length of the steam is recorded. This would determine the surface velocity of the stream. Several measurements are taken (about 4 or 5) and the average time noted gives a nearly correct surface velocity. The distance divided by the time gives the average surface velocity. The surface velocity is always higher than the average stream velocity. Therefore, the surface velocity is multiplied by a constant of 0.85 to get the average velocity. The cross-sectional area of the stream is calculated by multiplying the mean width of the wetted part of the channel by the depth of water. The average velocity multiplied by the average cross-sectional area gives the discharge

$$\text{Velocity} = \frac{\text{Length of channel (m)}}{\text{Average time taken by float (sec.)}} \times 0.85$$

2. Current meter method

Two types of current meters namely; the cup type and the propeller type are used to measure velocity in irrigation channels, streams or rivers. The cup type current meter consists of a wheel having several cups and the propeller type consists of vanes or propeller attached to a rod. When the meter is placed in flowing water, the velocity of water rotates the wheel or the vane. In case of shallow streams, the current meter is suspended from a rod and for measurement in deep water, a cable is used to suspend the meter. In large rivers, a boat is used to take readings inside the river and sometimes suspended from a bridge if available. As the wheel rotates, the revolution in a given period of time is recorded in a counter. A calibration chart or a graph gives the relationship between the revolution per minute and velocity. When mean velocity of a stream is determined with current meter, the cross section of flow is divided into a number of sub-areas and separate measurements are made for each sub area. The average velocity is obtained by taking the average of the velocities at 80% and 20% regions of the depth below the surface (from the top) in a vertical plane. In case of shallow stream, only one reading may be taken at 60% region of the depth from the surface. Discharge is estimated by multiplying the mean velocity of water by the cross-sectional area of the stream.

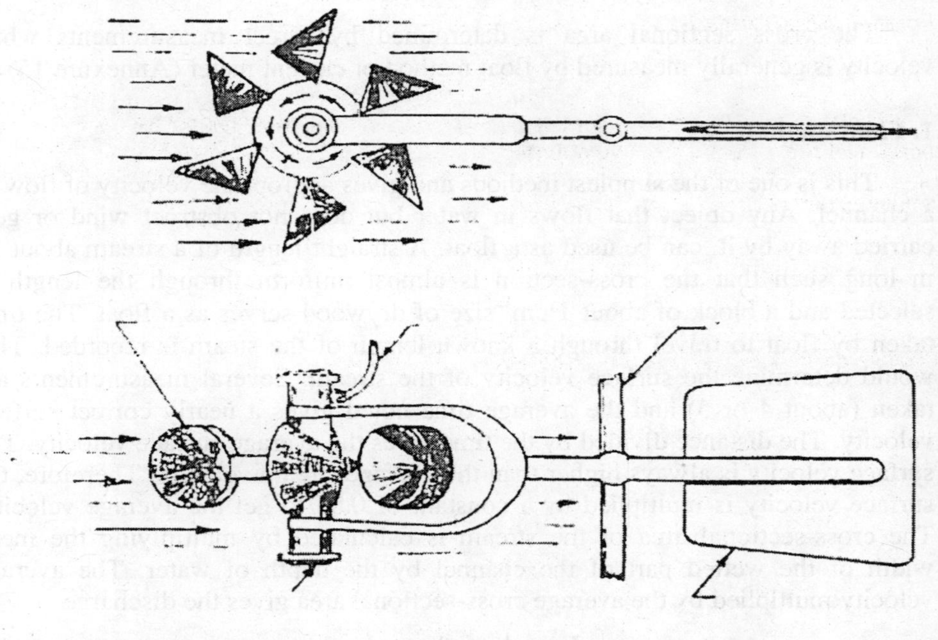

Fig 2. Different types of current meter (Courtesy: Rao, V.P. *et al.*, 2010)

3. Pipe method

This coordinate method is used to estimate discharges from flowing vertically or from pumping plants discharging horizontally. The method presents difficulties in accurately measuring the coordinates of the jet and therefore is of limited use.

(iii) Direct Measuring structures

In farm irrigation practice, the most commonly used devices for measuring water are weirs, parshall flumes, meter gate, orifices, siphons etc. In these devices, the rate of flow is measured directly by making a reading on a scale which is a part of the instrument and computing the discharge rate from standard formula. The discharge rate can also be obtained from ready reckoners prepared for the instrument.

1. Weirs

A weir is used to denote a notch or opening of some definite shape installed in a channel or a stream through which water falls. Materials like wood, concrete, mild steel, rigid PVC etc. can be used for construction of a weir but mainly a cut on a sheet metal is used for fabrication of weirs and installed at various locations in channel. Weirs can be classified as (i) broad crested and (ii) sharp crested. The sharp crested weir is mostly used for measurement of irrigation water. It is a weir with thin edge such that the sheet of water flowing over it has the minimum contact area with weir. Weirs may be rectangular, trapezoidal or triangular (90^0-V) but rectangular and 90^0V-notch weirs are commonly used on the farm. A weir when properly constructed and installed is one of the simplest and most accurate methods of measuring water.

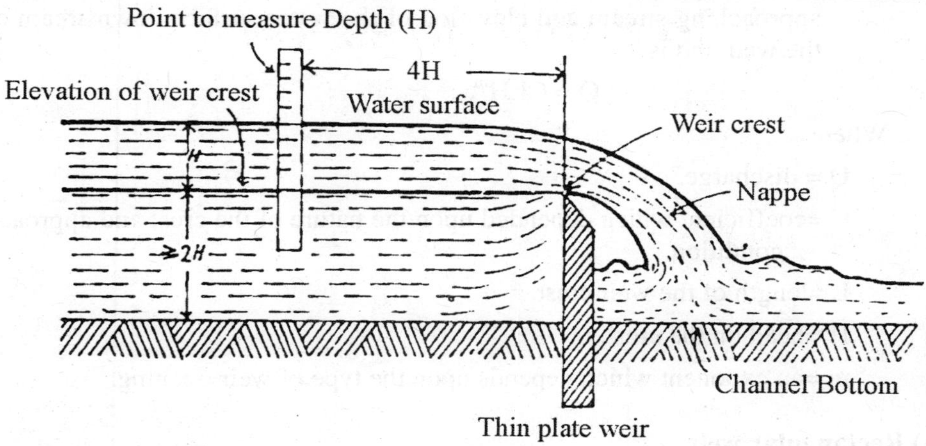

Fig 3. Installation of weir and its components (Courtesy: Rao, V.P. *et al.*, 2010)

Terminology

- The bottom most portion of the weir in touch with the water is called the *weir crest.*
- The sheet of water flowing over the weir is called the nappe.
- The top surface is the upper nappe and the bottom surface is the lower nappe.
- The depth of flow over the crest is known as head (H).
- It is measured at a minimum distance of 4 H upstream from the crest.
- The horizontal distance from the end of the crest to the side of the channel is known as end contraction.
- If both ends of the crest are away from the sides of the channel, then there are two end contractions.
- If the length of the crest is same as the width of the channel, then there is no end contraction (suppressed).
- The weir is said to have free flow condition, if the surface of water downstream is below the crest level so that the nappe is surrounded by air.

- If the downstream water level is higher than the weir crest level then it is a submerged flow condition.
- Apart from being classified as sharp crested and broad crested, weirs can also be classified according to the shape of the notch.
- The discharge through a weir is proportional to the head on the crest which depends on condition of the crest, contraction, velocity of approaching stream and elevation of the water surface downstream of the weir and is

$$Q = CLH^{m}.$$

Where

Q = discharge.

C =coefficient which depended upon the nature of the crest and approach condition.

L =length of the weir crest.

H =head on the crest.

m =an exponent which depends upon the type of weir opening.

(a) Rectangular weir

It is used to measure comparatively high discharges. Its crest is horizontal and the sides are vertical. The length of a rectangular weir may be equal to the width of the channel (suppressed rectangular weir) or less (contracted rectangular weir).

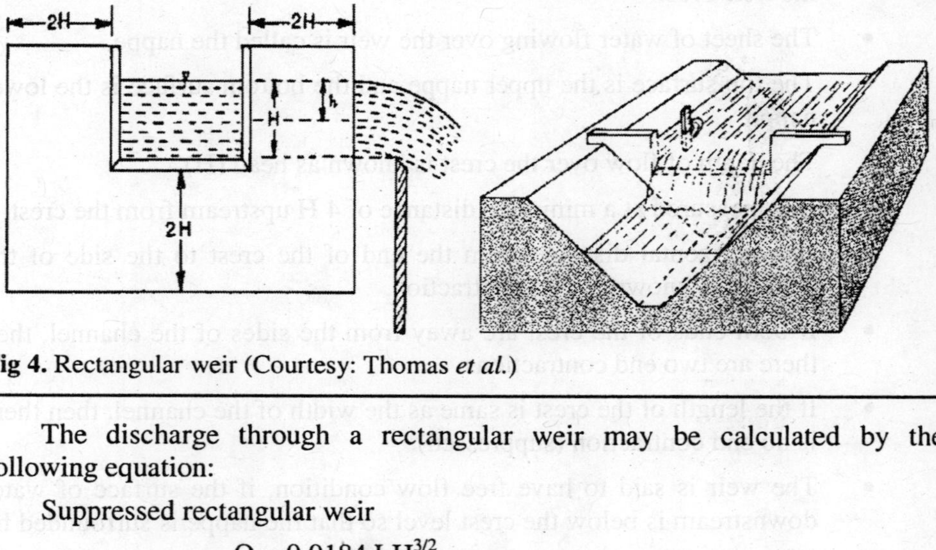

Fig 4. Rectangular weir (Courtesy: Thomas *et al.*)

The discharge through a rectangular weir may be calculated by the following equation:

Suppressed rectangular weir

$$Q = 0.0184\ LH^{3/2}$$

Where,

Q = Discharge (L/s)

L = Length of crest (cm)

H = Head over the crest (cm)

Contracted rectangular weir

$Q = 0.0184\ (L - 0.1\ nH)\ H^{3/2}$ Where,

Q = Discharge (L/s)

L = Length of crest (cm)

H = Head over the crest (cm)

n= number of end contractions

(b) Trapezoidal or cipolletti weir

The trapezoidal weir is a sharp crested weir and has a horizontal crest and the slides slope outward to give the notch a trapezoidal cross-section. Commonly a side slope of 1 horizontal to 4 vertical is used so it is named as cipoletti weir. It does not require any correction for end contractions and is used for measurement of medium discharges.

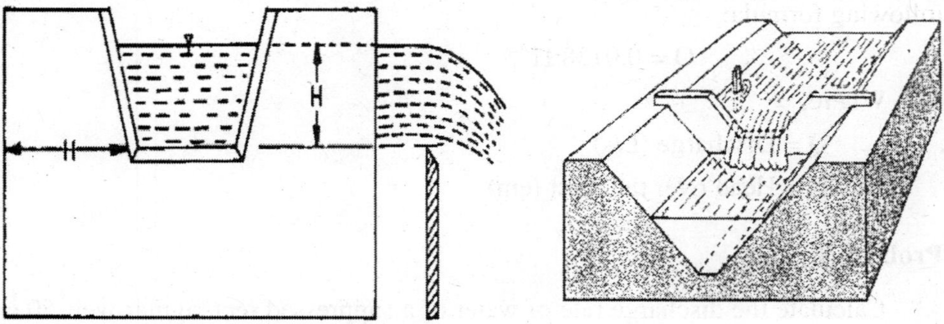

Fig. 5. Trapezoidal weir (Courtesy: Thomas *et al.*)

The discharge of water through this type of weir may be computed by the formula given below:

$$Q = 0.0186\ LH^{3/2}$$

Where,

Q = Discharge (L/s)

L = Length of crest (cm)

H = Head over the crest (cm)

(c) Triangular or 90° V-notch weir

It is used to measure small to medium discharges. The advantage of the V-notch weir is its ability to measure small flow accurately. The sides of the notch make an angle 45° with the vertical which gives a slope of 1 horizontal to 1 vertical (Annexure 16).

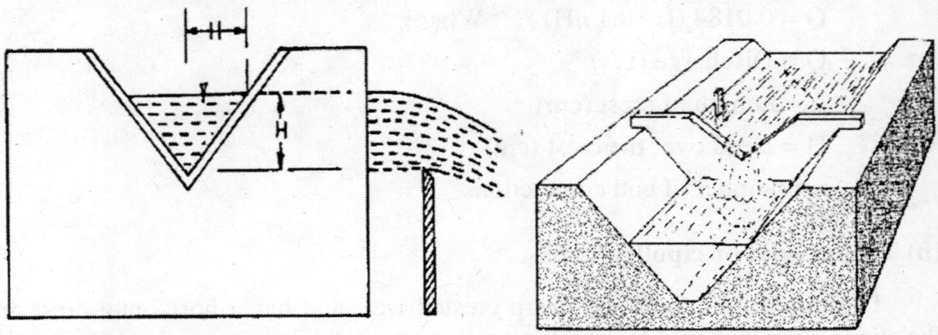

Fig 6. 'V'-Notch or Triangular weir (Courtesy: Thomas *et al.*)

The discharge through a 90^0V-notch weir may be computed by the following formula:

$$Q = 0.0138\ H^{5/2}$$

Where,

Q = Discharge (L/s)

H = Head over the crest (cm)

Problem 1

Calculate the discharge rate of water of a suppressed rectangular weir 50 cm long with a head of 12 cm.

Solution:

$$Q = 0.0184 \times 50 \times 12 \sqrt{12}$$
$$= 0.0184 \times 600 \times 3.46$$
$$= 38.2\ L/s$$

Problem 2

Calculate the discharge rate of water of a two end contracted rectangular weir 40 cm long with a head of 10 cm.

Solution :

$$
\begin{aligned}
Q &= 0.0184\ (L - 0.1n\ H)\ H^{3/2} \\
&= 0.0184\ (40 - 0.2 \times 10)\ 10\ \sqrt{10} \\
&= 0.0184 \times 38 \times 10 \times 3.16 \\
&= 0.0184 \times 380 \times 3.16 \\
&= 22.1\ L/s
\end{aligned}
$$

Problem 3

Calculate the discharge rate of water of a 90° V-notch with a head of 12 cm.

Discharge $Q = 0.0138 \times H^{5/2}$

$$
\begin{aligned}
Q &= 0.0138 \times 122 \times \sqrt{12} \\
&= 0.0138 \times 144 \times 3.46 \\
&= 6.9\ L/s
\end{aligned}
$$

Problem 4

A cipoletti weir of 50 cm crest length is installed in straight channel and the head over the crest of the weir is 15 cm. Calculate the discharge?

Solution:

$$
\begin{aligned}
Q &= 0.0186LH^{3/2} \\
&= 0.0186 \times 50 \times 15 \times \sqrt{15} \\
&= 0.0186 \times 50 \times 15 \times 3.87 \\
&= 54.03\ L/s
\end{aligned}
$$

2. Orifices

An orifice can be used to measure comparatively small streams like flow into border strips, furrows or check basins. An orifice is an opening in a bulk-head with closed perimeter usually circular (usually ranging from 2.5 cm to 7.5 cm in diameter) or rectangular cross-section through which water flows. An orifice is fabricated by making an accurate sharp edge cut of proper size and shape in iron, mild steel sheet, aluminum plate etc. through which water flows. The orifice flow formula can be applied only when the orifice flows fully, if it flows partially, it behaves like a weir. A plastic scale may be fixed directly on the upstream face of the orifice plate with its zero reading coinciding with centre of the orifice.

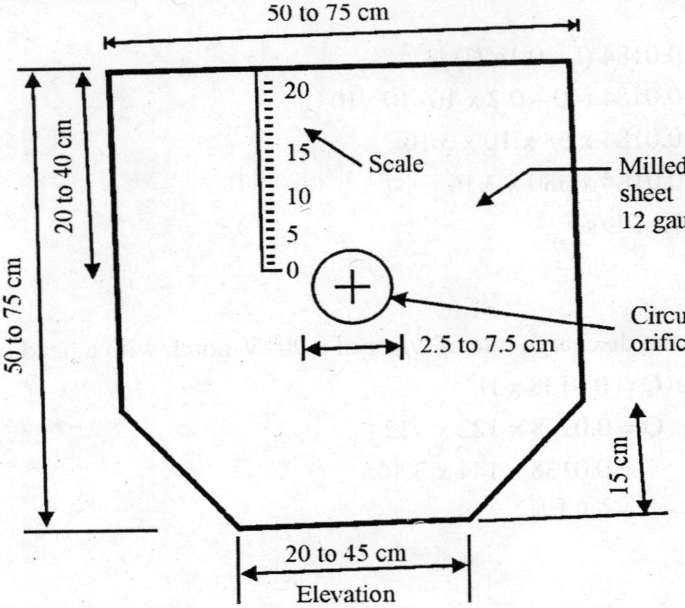

Fig 7. Orifice (Courtesy: Thomas *et al.*)

Free flow orifice: If the downstream water level is lower than the bottom of the orifice opening, the water jet discharges freely into the air and then it is said to have a free discharge. In case of free discharge, the depth of water on the upstream side of the orifice that causes the flow is called the head (h). Free flow orifices plates can be used to measure comparatively small streams like in border strips, furrows or check basins.

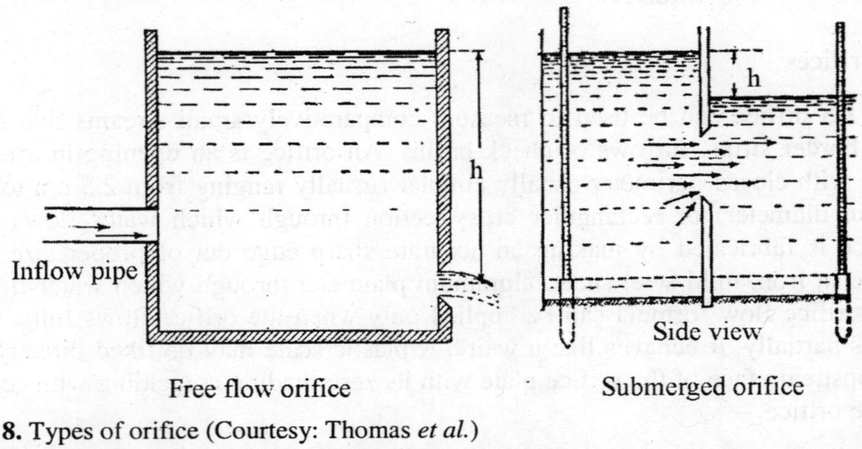

Fig 8. Types of orifice (Courtesy: Thomas *et al.*)

Submerged orifice: If the orifice discharges under a condition of submergence (the downstream water level is above the top of the opening and the flow discharges into water), it is called submerged. In case of submerged orifice, the difference of water level in the upstream and downstream sides is known as the head. No drop as in case of a weir is required for installation of orifices and therefore, this can be used more widely. The submerged orifice generally has a rectangular shape.

The discharge through an orifice is calculated by the formula:

$$Q = 0.61 \times 10^{-3} \times a \sqrt{2gH}$$

Where,

Q = Discharge through orifice, L/s

a = Area of cross section of the orifice, cm^2

g = Acceleration due to gravity, cm/sec^2 (981 cm/sec^2)

H = Depth (Head) of water over the centre of the orifice, cm.

Problem

A circular orifice has a diameter of 10 cm and the depth of water on the upstream side is 17 cm above the bottom of the orifice. If the coefficient of discharge is 0.61, calculate the discharge for free flow condition.

Solution: Data given

H = Head above the centre of the orifice = 17 -5 = 12 cm

$$a = \frac{\pi}{4} \times D^2 = \frac{\pi}{4} \times 10^2 = 78.54\,cm^2$$

The discharge through orifice can be calculated by the following formula:

$$Q = 0.61 \times 10^{-3} \times a \sqrt{2gH}$$

$$Q = 0.61 \times 10^{-3} \times 78.54 \times \sqrt{2 \times 981 \times 12} = 7.35 \text{ litre per second}$$

Advantages of weirs and orifice

- The accuracy in water measurement and it may vary from 80 to 85 per cent.
- Its simplicity and ease of construction make it more popular.

Disadvantages of weirs and orifice

- They require considerable head loss or fall of water surface making their use in area having land impracticable.
- They get silted up easily and the accuracy of the measurement is affected.

Flumes

Flumes are shaped, open channel flow sections that force flow to accelerate. Acceleration is produced by converging the side walls, raising the bottom or combination of both. Flumes range in size 2.5 cm to 15 m used in ditches and large canals. Parshall flumes and cut throat flumes are the most commonly used flumes for discharge measurement at the farms.

Parshall flume

Parshall (1950) designed this flume to measure the flow in which the discharge rate of water is obtained by measuring the drop in head or by allowing the stream of water to flow through a converged section of a flume with a depressed bottom. It is a self cleaning device. It is suitable for measuring a wide range of stream sizes, from a few L/s. to several cubic metres/sec. It can be used in relatively shallow channels (irrigation furrows) to large streams (canals). Here, the loss of head is very small hence flow condition is measured accurately to obtain the discharge. It can be constructed of G.I. sheet, aluminum sheet, M.S. sheet, brick masonry, reinforced concrete or timber. Flumes made of G.I. or aluminum sheets are easily portable. Brick masonry and concrete flumes are permanently constructed in the channels.

The Parshall flume has three main sections:

1. **Converging section at upstream end:** The floor of the upstream converging section is level with vertical walls converge towards the throat.
2. **Constricted section or throat:** The floor of the throat is inclined downward at a slope of 9 vertical to 24 horizontal, but the walls are parallel. The throat width specifies the dimensions of the parshall flume. The flumes of about 7.5, 15.0 and 22.5 cm sizes are used in field measurements.
3. **Diverging or expanding section at downward:** The floor of the diverging section slopes upwards and the walls diverge downstream.

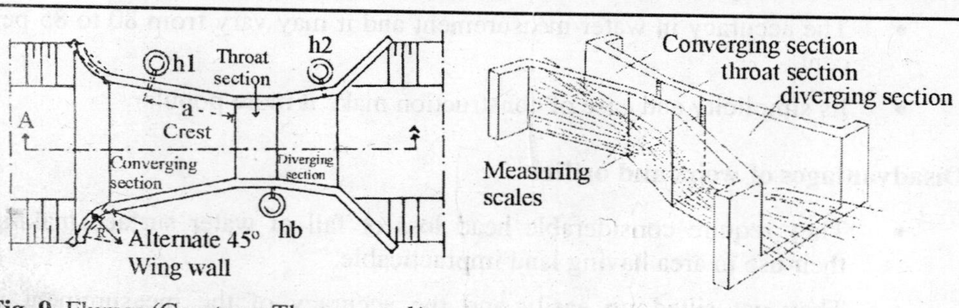

Fig, 9. Flumes (Courtesy: Thomas *et al.*)

Stilling wells should be constructed at the side to take measurements. A graduated scale is fixed in a stilling well constructed in the converging section (Ha) to measure the head of water over the floor of this section and to calculate the stream size. Another scale is fixed in the other stilling well at the beginning of the divergent section (Hb) to measure the head at that section for necessary correction for submerged flow. The size of flume is decided by the size of throat. Flumes are built of wood, steel sheet, concrete blocks or concrete and bricks. A heavy metal sheet may be used for a smaller flume. Standard dimensions of Parshall flumes and discharges can be referred in any standard book for irrigation and water engineering. Free flow discharge for Parshall flume is given in annexure 17. For successful operation, the flume should be of correct size carefully. The probable maximum and minimum flows and the allowable head are required to be estimated. Maximum allowable flow should be on the grade of the channel and the freeboard at the place where installed. The freeboard is the upward distance from the normal water surface in the channel to the top of banks. Flumes should have a free flow possible.

Discharge through the flume can occur under the following conditions

1. Free flow: When elevation of the water surface near the downstream end of the throat section is not high enough to cause any retardation of the flow due to a back flow of water is termed as free flow. Under this condition, only one measurement of the head or water level (Ha) in the converging section is required to estimate the rate of flow.

2. Submerged flow: When the elevation of the water surface downstream from the flume is high enough to retard the rate of discharge, the flow is said to be submerged flow. The important point in Parshall flumes is the ratio between the reading at Hb and Ha heads. This ratio should not exceed 0.6 for 7.5, 15.0 and 22.5 cm size flumes. If the ratio exceeds 0.6, it is termed as submerged flow. To determine this ratio, Ha and Hb are provided at the upstream and downstream sections of the flume. In this case a correction is done in free flow conditions to obtain the correct rate of flow.

Width of throat (cm)	Free flow limit, H_b/H_a
2.5 to 7.5	0.5
15.0 to 22.5	0.6
30.0 to 240.0	0.7

B. Cut-throat flume: The cut-throat flume is an improvement over the Parshall flume and has only two sections, *viz.,* converging and diverging section (no throat section). The flume has a flat bottom, vertical walls and a zero length throat section. It may have both free flow and submerged flow. Under free flow condition, critical depth of flow occurs near the flume neck or flume throat, which is the junction of the converging and diverging sections. The ratio of incoming flow depth (ha) to the flume length should be less than 0.4. The flume length may be from 45 cm to 3 m at neck and width may vary from 2.5 cm to 1.8 m. The flume should be a straight section of the channel. It is desirable to operate it under free flow conditions.

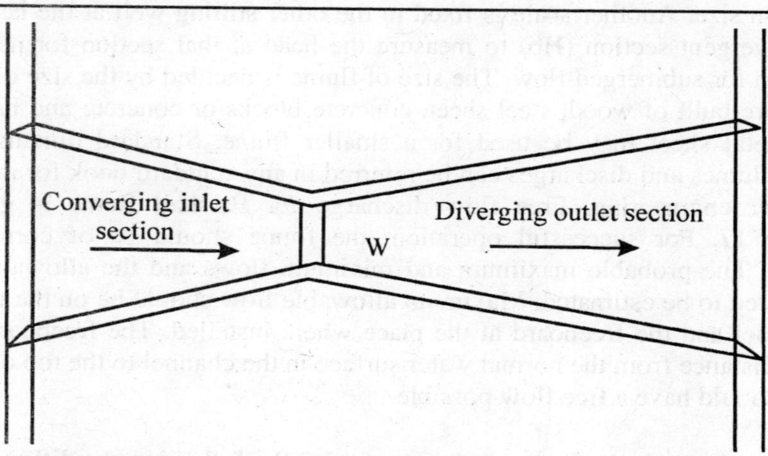

Fig. 10. Cut throat flumes (Courtesy: Thomas *et al.*)

Cut throat flume

The discharge through a cut throat flume under free flow conditions is:

$$Q = C_1 h_a n_1$$

Where,

Q = flow rate, ft/sec

C_1 = free flow coefficient, which is the value of Q when ha is 30 cm which is the slope of the free flow rating curve

h_a = head at the converging section, ft

n_1 = exponent whose value depends on flume length.

The value of n_1 is a constant for all cut throat flumes of the same length regardless of their throat width. The value of n_1 can be obtained for any flume length between 1.5 feet and 9 feet by plotting Q and h_a on logarithmic paper and reading the intercept and slope and the discharge may be $Q = 1.09\ ha^{1.825}$. When the flow depth exceeds the critical depth, the flow becomes submerged. For finding out the discharges, flow depths both at the converging section (ha) and the diverging sections (hb) are considered. The submergence, S is expressed as the ratio, often expressed in %, of the downstream depth (hb) the upstream depth (ha).

Advantages of parshall flume

- The parshall flume does not require any substantial drop in the channel bed for its installation.

- The flume does not get clogged due to suspended silt or floating debris.
- The parshall flume can be constructed in various sizes to measure a wide range of discharges.
- Partly submerged flumes also allow accurate measurement of discharge.

Disadvantages of parshall flume

- The parshall flume cannot be combined with a turnout from a canal or storage pond.
- It is more expensive.
- The construction is difficult.
- Only the small size made of sheet metal is portable.
- Right selection of the size and proper setting for precise measurements require some expertise.

Siphons

These are generally used to irrigate the field from canals and channels. Siphons are the pipes usually made of rubber, plastic or aluminium. The discharge formula for orifice can be used to measure the discharge rate of siphons.

Tracer method

This method involves releasing tracer like dye (potassium permanganate) or chemical or radioactive substance for measuring the velocity of water. Concentration of the chemical in flowing water at the downstream point is estimated by chemical analysis or Geiger counters or others for radio isotopes. The following applies to both open channel and closed conduit flow.

- Velocity-area tracer discharge equation
- Tracer-dilution discharge equation.

Procedure

- Install the flumes/weirs/orifice in the field.
- Turn the water on and fill the upstream portion of the channel
- Increase the discharge to its maximum, making sure that the discharge over the device is flowing freely or specify the condition of device.
- Record the height of water i.e Head.
- Calculate the discharge (L/s) using prescribed formula.

■■■■

26

MEASUREMENT OF SOIL MOISTURE WITH GRAVITATIONAL METHOD

Introduction

Water is retained by soil particles in the form of a thin film around it, and in numerous small pores of the soil matrix with forces, such as surface tension, capillarity, cohesion and adhesion. Under field conditions, water content of soil is not static and changes with time. These variations occur due to heterogeneity of soil, crop, climatic conditions etc. Thus, the water adsorption, storage and release depend upon the type of soil and the energy state of water. Water content in soil under certain standard conditions is referred to as soil moisture constants. Determination of soil moisture content helps in estimation of available water in the root zone and soil water potential and subsequent changes in physical and chemical properties of soil under the influence of change in water content. However, once the water retention curve of a particular soil is determined, it becomes very easy to know about the soil moisture content of soils at a particular energy or suction level. The soil moisture content is commonly expressed as the weight percentage i.e. grams of water associated with 100 grams of dry soil or in volume percentage i.e. volume of water to the volume of the soil in given sample. Broadly, the methods used for determination of the soil moisture content are classified as: (i) direct methods and (ii) indirect methods and expressed as below.

I. Direct methods: These methods are based on the principle of removal of water from the soil samples by evaporation, leaching or chemical reaction, and the amount of water removed or the change in weight of the soil due to moisture loss is estimated thermo- gravimetrically either through oven–drying or by volumetric method. It is the most accurate but cumbersome and time consuming.

II. Indirect methods: Indirect methods estimate amount of soil water through the properties of water affected by soil water content. Indirect methods are preferred over direct methods because they are nondestructive, less time consuming and permit more frequent observations, however, these methods require an initial calibration.

When cost, ease of use, and reliability are considered, tensiometers and electrical resistance blocks are usually the most practical devices for measuring soil-water but depend on their properly installation, maintenance, and calibration in field. The gravimetric method can be used to calibrate tensiometers and electrical resistance blocks on the farm.

The methods of soil moisture measurement are further classified as below:

- Quantitative Methods
  - o Gravimetric soil sampling - percent and volume
  - o Neutron scatter method – neutron probe (Rapid method)
  - o Di-electric constant method – TDR and FDR
- Qualitative Methods
  - o Tensiometer - works up to 0.85 bar
  - o Gypsum block- works up to 0 to 15 bars.
  - o Hand-feel method – faster method but is prone to error since it requires experience

	Classification	Methods
I	Direct method	Gravimetric method
II	Indirect methods	
	a. Geophysical methods	i. Tensiometer ii. Electrical Resistance (Gypsum blocks) iii. Electrical resistivity iv. Inductive electromagnetic technique
	b. Nuclear techniques	i. Neutron scattering ii. Gamma ray attenuation iii. Neutron radiography iv. Dual energy gamma sources v. Computer aided tomography (CAT scan) vi. Tritium tagging vii. Use of stable isotopes
	c. High frequency electrical techniques	i. Capacitance electrical technique ii. Portable dielectric probes (PDPs' surface) iii. Time domain reflectometry (TDR)
	d. Remote sensing method	iv. Visible and near infra-red v. Thermal infra-red vi. Micro wave-passive and active

Fig. 1. Schematic diagram of classification of methods used for soil moisture measurement

Gravimetric method

a. Weight basis

Principle: Soil sample is placed in hot air oven at 105ºC until all moisture present in soil sample gets lost to get a constant weight. The difference in weight of moist soil and oven dried sample gives the amount of moisture present in soil sample.

Materials: Sampling screw auger, moisture box, weighing balance and drying oven.

Procedure

- Weigh the empty moisture box (W_1)
- Take a sample of soil about 50 g with soil auger or sampling tube in air tight moisture box and cover it immediately with its lid to prevent the loss of moisture by evaporation.
- Cover the box with a cloth in the field to avoid heating due to insulation. Carry the samples to the laboratory.
- While sampling in a crop field, select the spot within a row between the two plants during early seedling stage as near the plant as possible (about 10 cm away from the plant). With the advance in age of crop, the sample may be drawn at a point midway between the two rows.
- Tipping off some soil near the lid, weigh the sample on a balance correct to two decimal places in gram (W2). If field balance is available, the moist soil sample can be weighed just after sampling in the field.
- Place it into an oven to dry the sample at 105ºC until constant weight is obtained normally for 24-48 hours.
- Cool the moisture box slowly to room temperature and weigh again.
- Record the weight of the dried sample (W3).
- Calculate the moisture percentage on weight basis by the formula:

$$\text{Moisture content (\%) (on weight basis)} = \frac{W_2 - W_1}{W_2 - W_1} \times 100$$

$$= \frac{(\text{Wet weight of soil} - \text{Dry weight of soil})}{\text{Dry weight of soil}} \times 100$$

Or

$$\text{Moisture content (\%) (on weight basis)} = \frac{\text{Weight of moisture}}{\text{Weight of oven dry soil}} \times 100$$

b. Volume basis

Principle: Volume of water present in the known volume of dry soil is the product of the moisture content on dry weight basis and the bulk density of the soil.

Materials: Sampling tube or a core sampler, moisture box, weighing balance and drying oven.

Procedure

- Take the soil sample with the help of core sampler of known volume and transfer it in pre weighed moisture box.
- Cover it immediately with its lid. Cover the box with a cloth in the field to avoid heating due to insulation. Carry the samples to the laboratory.
- Tipping off some soil near the lid, weigh the sample on a balance correct to two decimal places in gram (W1). If field balance is available, the moist soil sample can be weighed just after sampling in the field.
- Dry the sample in an oven at 105°C until constant weight is obtained.
- Cool the cans slowly to room temperature and weigh again.
- Record the weight of the dried sample (W2).
- Weigh the empty moisture box (W3).
- Calculate the moisture percentage on volume basis by the formula or volumetric moisture content (%) can also be estimated from the moisture content estimated on dry weight basis:

Method 1: Weigh the fresh weight of the moist sample and oven dry sample. Percentage soil moisture by volume is given by:

$$\text{Moisture content (\%) (on volume basis)} = \frac{W_1 - W_2}{V} \times 100$$

$$= \frac{\text{(Wet weight} - \text{Dry weight)}}{\text{Volume of core} \times \text{density of water}} \times 100$$

$$= \frac{\text{Volume of water}}{\text{Volume of dry soil}} \times 100$$

Since specific gravity (density) of water is 1, weight of water and volume of water are numerically equal.

Method 2: If we know the percentage soil moisture by weight and bulk density, then percentage soil moisture by volume is:

Moisture content (%) (on volume basis) = Percentage of moisture content by weight × bulk density × 100

c. Depth basis

Principle: Depth of water present in the known depth of soil is the product of the moisture content on volume basis and the respective depth of the soil profile.

Materials: Sampling tube or a core sampler, moisture box, weighing balance and drying oven.

Procedure

i. Take the soil sample with the help of core sampler of known volume (V) upto desired depth

ii. Follow the steps for determination of soil moisture as in weight and volumetric basis

iii. The moisture content on depth basis can be calculated by the formula

Amount of water or moisture content (%) (on depth basis) = Percentage of moisture content by weight × Bulk density × depth of soil (cm)

Or

= Percentage of moisture content by volume basis × depth of soil (cm)

Exercise

From the following data, calculate percentage soil moisture by weight, percentage soil moisture by volume, and amount of water on depth basis.

i. Weight of the moisture box with wet sample (W1) = 296 g

ii. Weight of the moisture box with dry sample (W2)= 246 g

iii. Weight of the empty moisture box (W3)= 36 g

iv. Bulk density soil= 1.38 g/cm^3

v. Depth of soil profile= 45 cm

Solution

Weight of water in the sample= 296 – 246= 50g

Weight of oven dry soil= 246 – 360= 210 g

$$\text{Moisture content (\%) (on weight basis)} = \frac{W_1 - W_2}{W_2 - W_3} \times 100 = \frac{50}{100} \times 100 = 23.8\%$$

Moisture content (%) (on volume basis)

= Percentage of moisture content by weight × Bulk density

Moisture content (%) (on volume basis) = 23.8 × 1.38 = 32.8%

Moisture content (%) (on depth basis) = Percentage of moisture content by volume basis × depth of soil (cm) = $32.8\times\frac{45}{100}=14.76\,\text{cm}$

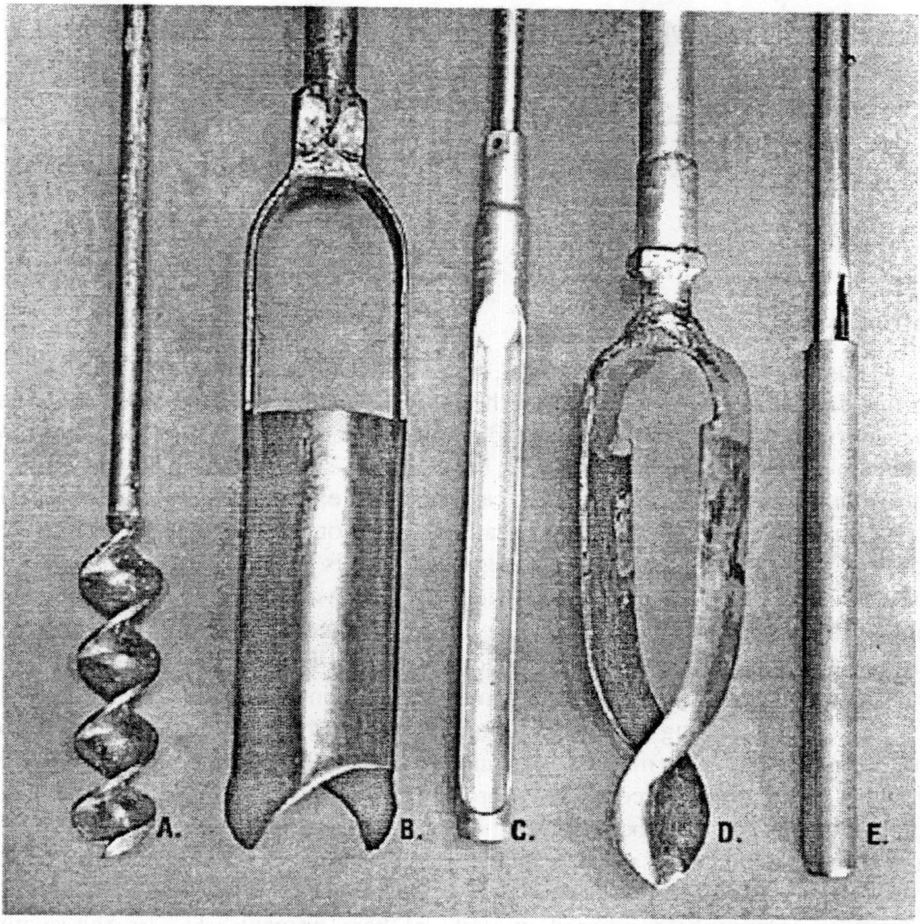

Fig. 2. Soil augers for soil sampling: A – Screw auger, B – Barrel auger, C – Sampling tube, D – Dutch mud auger and E – peat sampler. (Courtesy: Rao *et al.*, 2010(b)).

27

MEASUREMENT OF SOIL MOISTURE WITH TENSIOMETER

Introduction

Tensiometer is a sealed, airtight, water-filled tube (barrel) with a porous tip on one end and a vacuum gauge on the other, measures the tension or soil water suction (negative pressure) with which water is held in soils and the amount of water in the soil that is available to plants in-situ rapidly. This suction is equivalent to the force or energy that a plant must exert to extract water from the soil per unit area. The tensiometer read as matric or capillary potential only and does not read osmotic potential, as the porous cup is permeable to both water and solutes. After stabilization, water inside the tube assumes same composition as that soil water. Here, the water in the tensiometer equilibrates through a porous cup with adjacent soil water and suction in the soil is same as the suction in the tensiometer. The water moves out or in the cup depending upon the tension of the soil water. Scheduling irriagations based on tensiometric reading maximizes irrigation efficiency by minimizing runoff and percolation losses (Optimize crop yields). The suction force in the porous tip is transmitted through the water column inside the tube because of the cohesive forces between adjacent water molecules and displayed as a tension reading on the vacuum gauge or manometer with suction scale of 0 to 100 cb as on most commercial tensiometers. Soil-water tension is commonly expressed in units of bars or centibars. As the suction approaches approximately 0.8 bar (80 cb), the cohesive forces are exceeded by the suction and the water molecules separate. When this occurs, air can enter the tube through the porous tip and the tensiometer no longer functions correctly. This condition is referred to as *breaking tension*. A curve for each soil is needed to determine the moisture percentage. The range of its usefulness is 0 to 0.8 bar suction and suitable for sandy, loamy sand, sandy loam, and the coarser-textured of loam and sandy clay loam soils. It can measure soil moisture at different depth by placing porous cup fitted at the end of the tensiometer at desired depths (Annexure 18).

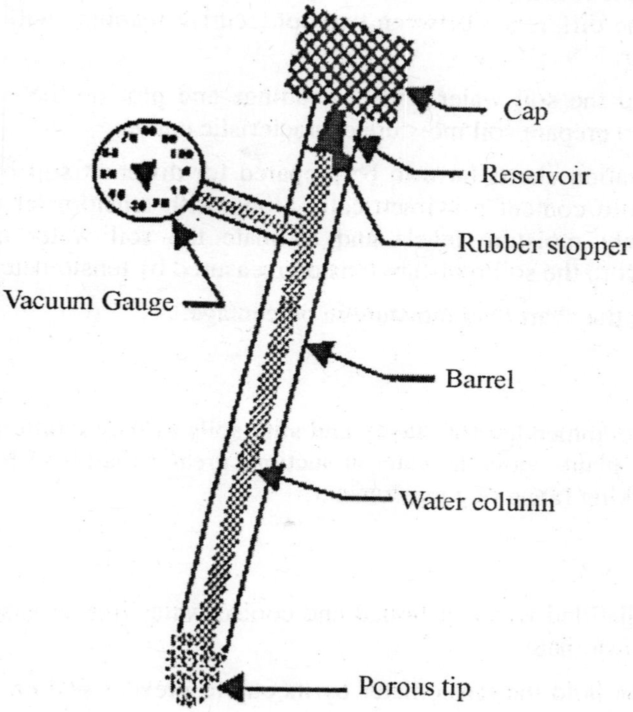

Fig. 1. Tensiometer

Procedure of measurement

- Select a representative spot in the cropped field, usually between plants within a row.
- Make a vertical hole equal to the cup diameter by driving a hollow pipe with sharp cutting edges or using a screw auger to the desired depth and remove it carefully with a precaution that no soil falls back into the hole.
- Fill the tensiometer with distilled water and observe for any air bubbles in the tube.
- Insert the tensiometer into the hole by pushing directly down on the cap so that the walls of the cup are in firm contact with the soil. Compact the soil around the stem of the tensiometer by adding some loose soil near the surface and make a small heap so that water will not collect near the tensiometer and seep down along its body.
- Allow the tension meter to attain equilibrium and start monitoring the soil water tension reading in the vacuum gauge.

- Take the readings in the morning and record the readings frequently so that the difference between two consecutive readings will be less than 0.1 bar.
- Record the soil water tension readings and plot on the graph against days to prepare soil moisture characteristic curve.
- Calibration charts have to be prepared for different soil by measuring moisture content gravimetrically along with tensiometer readings for different moisture levels and estimate the soil water content with respect to the soil moisture tension measured by tensiometer.
- Using the chart read moisture in percentage.

Limitations

It is not recommended for clayey and silty soils as they retain more than 50 percent of their plant-available water at suctions greater than 0.85 bars, which is outside the working range of a tensiometer.

Precautions

- Use distilled water or boiled and cooled water free of gases for filling the tensiometer.
- Do not hold the tensiometer by its cup to prevent sealing of pores due to greasy palms.
- Do not rotate the tensiometer shaft and do not push on the guage.
- When installing tensiometers, the tips must be in the active root zone of the crops grown.
- The instrument must be installed properly so that the porous tip is in good contact with the soil, ensuring that the soil-water suction is in equilibrium with the water suction in the tip.
- Install the device with reasonable care, and it must be protected from damages from livestock, cultivation and periods of freezing temperatures.
- Water should be drained from the instrument and the cup, and washed with clean water before storage after removing from soil.

■■■■

28

MEASUREMENT OF SOIL MOISTURE WITH ELECTRICAL RESISTANCE METHOD

INTRODUCTION

The electrical resistance or gypsum block method (Bouyoucos and Mick, 1940) is the most widely used for indirect measurement of soil moisture in relatively all types of soils. It consists of two electrodes enclosed in a block of porous material made of gypsum, fiberglass or nylon. Gypsum block is found to be less sensitive to soil salts than nylon and fiber glass. The electrodes are connected to insulated lead wires that extend upward to the soil surface. When properly installed, the water suction of the porous block is in equilibrium with the soil-water suction of the surrounding soil. As the soil moisture changes, the water content of the porous block also changes. The electrical resistance between the two electrodes increases as the water content of the porous block decreases. The resistance readings (soil water reading) are taken with a portable A.C. wheatstone bridge operated with dry cells reading from 0 to 100 or 0 to 200 voltage normally. High readings on the scale (corresponding to low electrical resistance) indicate high levels of soil-water, whereas low meter readings indicate low moisture levels. It works on the principle that water conducts electrical conductivity when two electrodes are placed parallel to each other in a porous non-conducting medium. The resistance to the flow of electricity is inversely proportional to the moisture content in the medium when electrical current is passed. The actual resistance reading of a block varies with the type of electrodes, their length and distance between them; matrix material, its density and temperature. It is necessary to apply a temperature correction to the block readings.

This method gives reasonably accurate moisture reading over the range of 1 to 15 atmosphere suction. Therefore, resistance blocks are best suited for use in fine textured soils such as silts and clays that retain at least 50 percent of their plant available water at suctions greater than 0.5 bar. The resistance in conductivity is determined by the moisture content and in turn by the tension or suction of water in the nearby soil. The relationship between the resistance reading and the soil moisture percentage can be determined by calibration. A widely used material to make the blocks is gypsum. Gypsum blocks are generally made in rectangular shape. The size is usually 5.0 cm × 4.0 cm × 2.5 cm. The pair of electrodes is usually made of 22 mesh stainless steel wire screen soldered to the copper lead wire. Typically, the screen electrodes are 1.25 cm long and 0.5 cm broad connected to 0.1 cm diameter thick single core PVC coated wire. The electrodes are spaced at 1.0 cm apart and embedded at the centre of the block. Gypsum blocks are prepared using 10 parts of plaster of paris and 8 parts of water. It sets in about half hour. The pore size of the material used in most electrical resistance blocks does not permit changes in water content dramatically and accordingly electrical resistance of the block.

Procedure

- Select a convenient spot for installation normally within row in between plants.
- Sink a bore with a post-hole auger or soil probe to the depth of installation of blocks. The depth of bore varies with crop root zone.
- Pour about a little water (about half cup) into the hole to form slurry of mud in the bottom.
- By running the lead wires through the probe tip and holding the block firmly against the tip with wire, push the block firmly to the bottom of the hole, forcing the slurry to envelop the block.
- Place the blocks at desired depth in the soil, then fill back the bore in small depths by trampling the soil with a wooden rod in its natural order.
- After placement ensure that there is a firm contact of the blocks with the soil.
- Make a separate hole for each block to desired depth.
- Heap the soil to a height of about 3 cm near the surface at bore spot to prevent any water stagnation.
- Irrigate the field and record the soil moisture depletion or electrical resistance developed at the electrodes by resistance meter or using a wheatstone bridge.
- This resistance is interpreted in terms of gravimetric soil moisture by using calibration charts.

Calibration of instrument

- Resistance blocks are generally calibrated in terms of soil water tension to make readings applicable across soil textures.
- Blocks should be calibrated for each soil type. Often, calibration curves are provided by manufacturers or manually after installing it in field. First of all irrigate the field and as the soil dries, record the resistance reading periodically with the wheatstone bridge and also determine the

soil moisture content of the sample taken from the block depths gravimetrically.

- Plot the data of resistance reading and soil moisture percentage during the drying cycle, on a graph paper. Record the process two times and draw the curve, as these resistance units can be used for direct measurement of soil moisture tension.

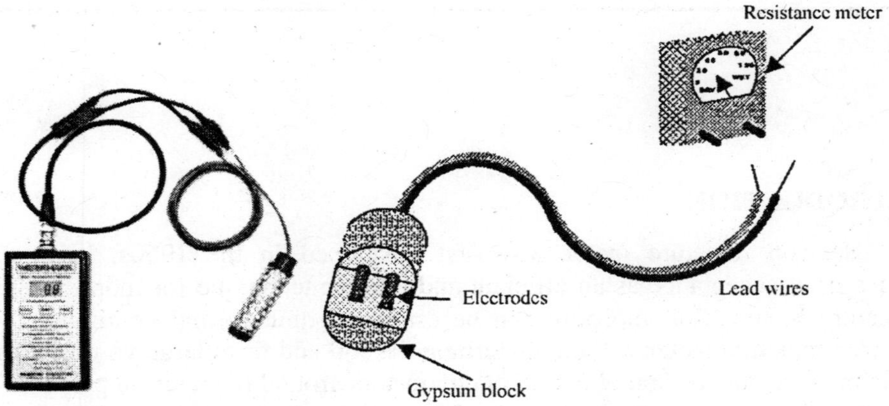

Fig. 1. Gypsum blocks and moisture meter

Limitations

- In the wetter regime (0 to –0.8 bar), the sensitivity of electrical resistance meters is not so accurate compared to tensiometers, thus the resistance blocks characterise potentials less than about –0.85 bar.
- Electrical resistance blocks are not reliable for determining as when to irrigate sandy soils where over 50 percent of the plant-available water is usually depleted at suctions less than 0.5 bar.

Precautions

- The blocks must have close contact with undisturbed soil.
- There should not be any root pieces, pebbles etc.
- When installing in a crop, ensure that these are placed in the root zone.
- Do not place the two blocks at vertical interval less than 30 cm.
- Standardize the blocks before installation to eliminate trapped air in the block. It was done by soaking it in water because soaking removes air from the blocks and insures accurate meter readings as per instructions before installation.

29

MEASUREMENT OF SOIL MOISTURE WITH NEUTRON MOISTURE METER

INTRODUCTION

Neutron moisture meter was first developed in the 1950s, has gained widespread acceptance as an efficient and reliable technique for monitoring soil moisture in-situ. Soil moisture can be estimated quickly and continually with neutron moisture meter without disturbing the soil and from large volume of soil. This meter scans the soil about 15 cm diameters around the neutron probe in wet soil and 50 cm in dry soil. In the soil, if fast neutrons are emitted in all directions, they collide repeatedly with atoms of water in soil. Through repeated collisions, they are scattered and lose some part of their energy. These slowed neutrons are said to be thermalized, and called "slow neutrons." The neutron detector in the neutron moisture meter is designed to ignore fast neutrons and count only thermalized or slow neutrons. Often, the source of fast neutrons is obtained by mixing alpha particles from a radioactive emitter with beryllium. The fast neutrons (having an energy range of 2 – 4 MeV and an average speed of about 1600 km/sec) are emitted radially into the soil.

The average energy in a collision is much higher with atoms having low atomic weight as compared to heavier atoms. In typical soils, hydrogen is the only element with a low atomic weight occurring in significant amounts. As the major source of H^+ atoms in the soil is water molecules, the number of neutrons slowed due to collision is directly proportional to the moisture content of the soil. This property is made use of in a neutron moisture meter, where in, fast moving neutrons from a radioactive source are allowed to hit on the H^+ atoms of water present in the soil, and the number of slowed down neutrons because of the collision are found out. The density of the cloud of thermal neutrons is a function of water content in the soil. The density of the neutron cloud is measured by the detector and the resultant signals are processed electronically and displayed as a number on the gauge. The gauge reading is an index of soil moisture. This is best method (at present available methods) for quantities estimation of changes in soil moisture.

Moisture content of the soil can be determined from the calibration curve with count rate of slow neutrons of that particular soil against standard gravitational method. Soil moisture content is expressed as percentage by referring to calibration curve prepared by thermo-gravimetric methods.

Components of a neutron moisture meter: The major components of a neutron moisture meter are the following:

Probe: The device essentially consists of a probe made of stainless steel cylinder with 3.8 cm diameter and 75cm length. The probe is held inside a probe carrier made of PVC pipe. The probe contains a fast neutron source, which emits gamma radiation, and a detector of slow neutrons.

Source: The source of fast neutrons is obtained by mixing alpha particles from a radioactive emitter (either radium or americium) with beryllium.

Detector: The most commonly used detector to sense the density of thermal neutrons ignoring the fast neutrons and other radiations is lithium enriched boron trifluoride (BF_3) in the probe.

Rate scaler: It is a short counting unit attached through hinges to the upper end of the carrier. The scalar is battery operated and monitors the flux of slow neutrons. The rate scaler is connected to the cable by a socket provided at its base.

Procedure

- The access tube is inserted vertically into the soil upto the desired depth using an augur and forcing it to the drilled depth gently by a hammering.
- The gap between the access tube and the soil is filled with the slurry of soil.
- The fast neutrons in the probe are emitted radially into the soil.
- The neutrons emitted from the probe collide with or deflect from hydrogen ions (of which water is a major source) and are slowed.
- The density of the cloud of thermal neutrons generated within the soil around the probe is a function of water content in the soil. The density of thermal neutrons is sensed by BF_3 in the probe. For making the readings, the scaler is warmed up for about 20 to 30 minutes.
- The scalar monitors the flux of slow neutrons, which can be related to moisture content.
- Standard count is taken with the probe fully shielded.
- The counting time will depend upon the precision required.
- A neutron moisture meter must be calibrated to measure the actual soil water content for a particular soil type.

- Soil moisture content is expressed as percentage by referring to calibration curve prepared by thermo-gravimetric methods.

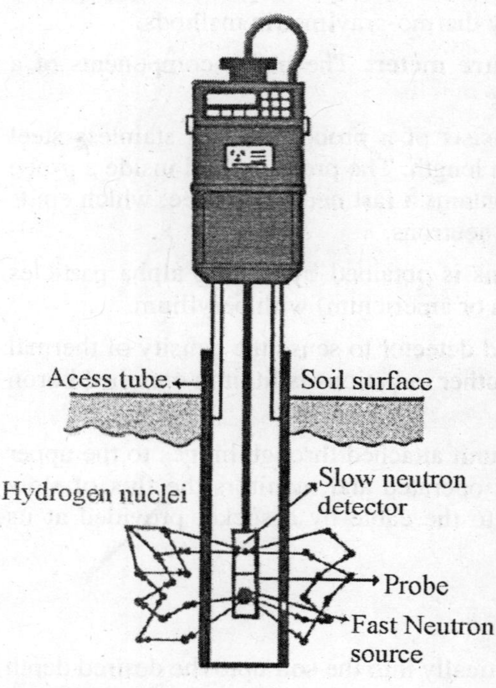

Fig. 1. Measurement of soil moisture by Neutron moisture meter (Courtesy: Rao *et al.* 2010 a).

Precautions

- By understanding the hazards, follow safeguard against possible injury during handling.

Merits

- Less laborious and time saving (2 to 5 minutes are enough for a single measurement)
- Non-destructive method after initial installation
- Repeated measurements can be made at the same location and depth with an accuracy of 1.0 % moisture content.
- Independent of temperature and pressure.
- Soil moisture can be estimated for long-term studies from large volume of soil.

- Its simplicity in operation

Demerits (Limitations)

- High initial and maintenance cost of the equipment.
- It has delicate electronic parts that may be difficult to maintain.
- Probe and access tube must be calibrated for each soil.
- Difficult to measure the soil moisture in the top 15 cm soil depth (surface soil or shallow soils) due to escape of neutrons into atmosphere.
- It is not suitable for changing water front in a soil.
- Health hazards due to exposure to neutron and gamma radiation
- Presence of Li, B, Cd or other nuclei of high neutron capture or hydrogen from organic matter may cause some inaccuracies in readings.
- Other forms of soil hydrogen, such as humus, plant organic matter, and water bound in the crystalline structure of calcium carbonate and gypsum may also cause inaccurate readings.

■■■■

30

MEASUREMENT OF SOIL MOISTURE WITH PRESSURE PLATE APPARATUS

Introduction

Pressure plate apparatus helps in determining the soil moisture content and water release pattern by desorption method over a wide range of suction pressures from field capacity to permanent wilting point (0.1 to - 15 bars). This equipment actually establishes the relationship between water content and metric potential and enables to develop soil moisture characteristic curves over a wide range of matric potential. Laboratory measurements of soil water potential are usually made with this apparatus. It consists of an air tight metallic chamber in which porous ceramic pressure plate is placed of high air entry values strong enough to withstand high pressure (-15 bars). The maximum air pressure is determined by the diameter of the largest pore.

The main objectives are to determine the moisture content of soil at FC and PWP to find out the capillary conductivity of soil and to prepare a water release/retention curve.

Materials and equipment

1. Pressure plate assembly (Pressure supply units, manifold and pressure plate chamber)
2. Soil sample
3. Moisture box
4. Empty wash bottle
5. Pipette
6. Filter paper
7. Pencil
8. Oven
9. Desiccators

Procedure

- Check the leakage of joints and air-entry valve of the ceramic plate to ensure that it is higher than the pressure to be applied.
- Saturate the sample, at least, for 24 hours with an excess of water in the ceramic plate.
- Remove the excess water with the help of pipette or an empty wash bottle.
- Pour all soil in the sample retaining rings at a time, to avoid particle size segregation.
- Put a piece of filter paper marked with sample identity number with pencil on the filter paper piece.
- The chamber is closed with special wrenches to tighten the nuts and bolts with required torque for sealing it through rubber seal or O-ring at its proper place.
- The required pressure of 0.33 or 15 bar is applied through a compressor and maintained at a desired level.
- Attach a burette to the outflow tube or the ceramic plate. Water starts to flow out from saturated soil samples through outlet and continues to trickle till equilibrium against the applied pressure is achieved.
- No change in the burette reading will indicate that the outflow has ceased. In general, equilibrium is attained within 48 hrs, if the height of the sample is 1 cm.
- Remove the burette from the outflow tube.
- Disconnect the source of air supply to the chamber by turning the regulator in an anti-clockwise direction.
- Release the pressure from the chamber.
- Open the chamber, and transfer the sample into already weighed can box.
- Weigh the can box with sample, and keep it in the oven at 105°C.
- Transfer the can box into a desiccator containing $CaCl_2$, and allow it to cool.
- Take the weight of the can box with oven-dry soil.
- The pairs of pressure and moisture content data so obtained are used to prepare soil moisture characteristic curve.

Precautions

- All the joints should be air-tight.
- It should be ensured that there is no leakage from the chamber.

- The pressure membrane, normally, should not be used more than once, at least, at higher pressure values (more than 10 atm.).
- Before storing a ceramic plate, it should be dried for several days (until dry) by covering the surface of the ceramic plate with a thin layer of fine dry soil. This will minimize the evaporation deposits on the surface of the ceramic plate.

Observations and Calculations

Sample No.

Particulars	**Tension Applied (bar)**					
	0.1	**0.2**	**1.0**	**2.0**	**5.0**	**15.0**
a. Wt. of empty box:						
b. Wt. of box + wet soil:						
c. Wt. of box + dry soil:						
d. Wt. of water (b-c):						
e. Wt. of dry soil (c-a)						
f. Soil water percentage [d/c x 100]						

Graphical evaluation of tension-water content (h-0) relation of the soil. The soil moisture characteristic curve is `L' shaped for sandy soils while `I' shape for clay soil (straight line)

Result

The soil moisture content at a given moisture tension is ---------------- %.

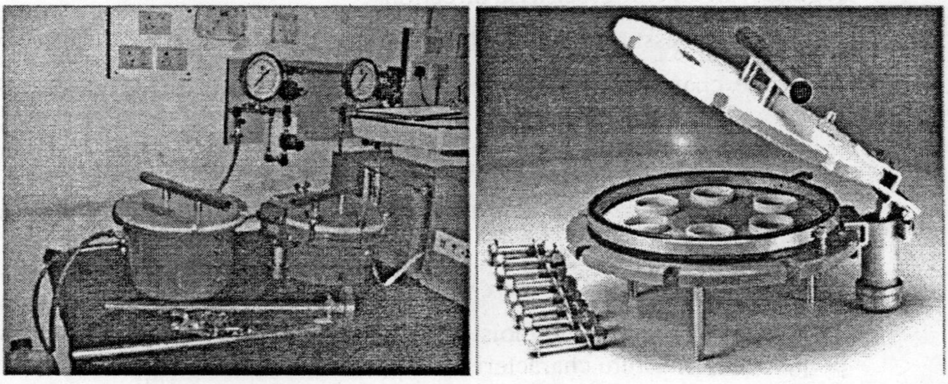

Fig: Pressure plate apparatus (Courtesy: Rao *et al.*, 2010a)

■■■■

31

SOIL MOISTURE MEASUREMENT BY TIME DOMAIN REFLECTOMETER

Introduction

The time domain reflectometer (TDR) is a new device developed to measure soil-water content. Two parallel rods or stiff wires are inserted into the soil to the depth at which the average water content is desired. The rods are connected to an instrument that sends an electromagnetic pulse (or wave) of energy along the rods. The rate at which the wave of energy is conducted into the soil and reflected back to the soil surface is directly related to the average water content of the soil. One instrument can be used for hundreds of pairs of rods. This device, just becoming commercially available, is easy to use and reliable.

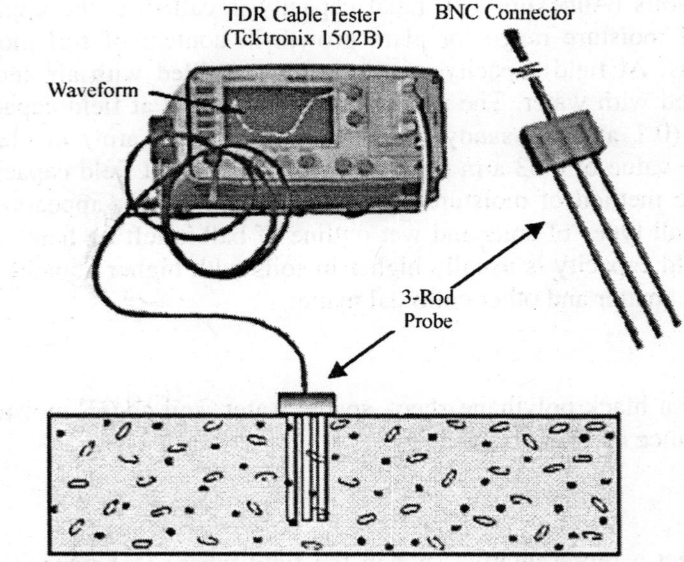

Fig. 1. TDR cable tester with three-rod probe embedded vertically in surface soil layer.

32

MEASUREMENT OF FIELD CAPACITY BY FIELD METHOD

INTRODUCTION

Field capacity is defined as amount of water held by soil against gravity on oven dry basis. It can be measured after complete saturation of soils after draining excess water. At this stage, the rate of downward movement is materially decreased to negligible value. This stage is attained generally 24-72 hrs. after saturation. At this stage, water is retained under the influence of metric suction. Here, the gravity becomes equal in magnitude of metric suction or tension. The measurement of field capacity is important for efficient water management and irrigation scheduling. Sandy soils reach field capacity earlier than clayey soils (Annexure 19). Field capacity is called as the upper limit of available soil moisture range for plant growth in context of soil moisture and plant relations. At field capacity, macro pores are filled with air and capillary poles are filled with water. The average matric potential at field capacity varies from 10kPa (0.1 atm) in sandy soils to 33 KPa (0.33 atm) in clayey soils. However, the value of 0.33 atm is commonly accepted. At field capacity, in feel or appearance method of moisture estimation, no free water appears on soil on squeezing in all types of soils and wet outline of ball is left on hand. Soil water content at field capacity is usually higher in soils with higher content of silt and clays, organic matter and other colloidal matters.

Materials

Straw or a black polythene sheet, spade, water, soil auger, moisture boxes, weighing balance and hot air oven.

Procedure

- Select a representative spot in the field where that water table is not within two metres.
- Bund an area of about 2 m^2 on all four sides and remove all weeds to avoid transpiration losses.

- Pond the area with water till the desired depths get sufficiently wet.
- Spread straw mulch of at least 30 cm thickness on the surface to prevent evaporation. The black polythene sheet can be conveniently used.
- Take soil sample at every 24 hours interval from the desired layer with a soil auger for 5-6 days till the two successive values of water content after drying in an oven at 105°C are nearly equal.
- Mark the readings on a graph having time interval and moisture content on oven-dry weight basis.
- The lowered influx value can be taken as the field capacity.

Precautions

- Sufficient water is ponded over the area to ensure that the desired soil layers get fully saturated.
- Soil surface should be clean (free from evaporative and transpiration agent i.e. weeds)

Calculation and Observations

$$\text{Field capacity (FC) (\% dry weight)} = \frac{W_2 - W_3}{W_3 - W_1} \times 100$$

1. Weight of empty moisture box (W_1) = g
2. Weight of moisture box + moist soil sample (W_2) = g
3. Weight of moisture box + oven dry soil (W_3) = g

Example

The moisture percentage of a sandy loam soil at 24 hours interval after saturation is as follows:

24 hours after saturation: 36.5%

48 hours after saturation: 29.5%

72 hours after saturation: 24.3%

96 hours after saturation: 24.1%

Results

Field capacity of above soil is 24.3%

■■■■

33

MEASUREMENT OF BULK DENSITY

INTRODUCTION

Bulk density refers to the mass of soil per unit volume of undisturbed soil including pore spaces, minerals and organic material. It is soil physical indicator and also called apparent specific gravity. It gives useful information in assessing the potential for leaching of nutrients, erosion, and soil productivity. Knowledge of bulk density is of particular importance in water management as it may often be required to express water content of the soil on volumetric basis (water storage capacity). Bulk densities limit plant growth and varies from soil to soil i.e. for different textural classes and nature of soil solids as well as the pore spaces (Table 1). The volume of soil is not always constant as some clays swell when wet and shrink when dry. To avoid such errors, usually bulk density is measured at field capacity. The bulk density may be expressed as Mega gram per cubic meter (Mg m^{-3})/gram per cubic centimeter (g cm^{-3})/pound per cubic feet (lb $feet^{-3}$). The value for bulk density (ρb) of a soil is always lower than its particle density. Its value ranges from 1.0 to 1.5 g/cm^3. For most cultivated soils, it is around 1.3 g/cm^3. Bulk density, in general measured by following methods.

(1) Core sampler method
(2) Weighing bottle method
(3) Clod coating method
(4) Clod saturation method

Table 1. Value of the bulk density for different soils

Particulars	Bulk density (Mg m^{-3})
Sand dominated soils	1.70
Organic peat soils	0.50
Compacted sub soils	2.00
Clay soil	1.10
Ideal soil	1.33
Normal soil	1.10 – 1.60

1. Core sampler method

Materials required: Core sampler, hot air oven, moisture cans and weighing balance.

Procedure

- Saturate the soil with water and cover it with some mulching material to reduce the loss of water by evaporation.
- Allow the soil to reach field capacity usually for 24 hours.
- Take a core sampler which has a cutting edge and drive it into soil at required depth.
- Dig out the soil surrounding the cylinder without disturbing the soil core.
- Weigh the soil contained in the core sampler and record the fresh weight.
- Collect the soil from the core sampler in a moisture box after trimming the sample at both ends of the core cylinders for determination of moisture percentage on oven dry basis.
- Record the weight and dry the soil in an oven at 105°C for about 24 hours or until all the moisture in the sample is driven off and till constant weight is obtained.
- Take the weight of empty box also.
- Calculate the volume of core sampler ($\pi r^2 h$) by finding out the internal diameter and height using vernier callipers as following:

  Volume of soil =Volume of core= $\pi r^2 h$

  Where,

  h =height of core sampler (cm) and

  r = radius of the core sampler (cm)
- Determine the bulk density of soil contained in the core sampler by using the formula.

$$\text{Bulk density (Bd)} = \frac{\text{Weight of oven dry soil (gram)}}{\text{Volume of soil (cm}^3\text{)}}$$

Precautions: No compaction should occur during driving the core so that a representative volume of soil having field structure is obtained.

Observations

Calculate the bulk density of soil using following observations from two soil samples assuming the radius (r) and height (h) of core as 2 and 4 cm, respectively.

Sample	Weight of empty box W_1 (g)	Weight of box + moist soil W_2 (g)	Weight of box + dry soil W_3 (g)
A	52	180.5	112
B	45	169.6	107

Calculation

Volume = $\pi r^2 h = 3.14 \times (2)^2 \times 4 = 50.24\ cm^3$

$$\text{Bulk density (Bd)} = \frac{\text{Weight of oven dry soil (gram)}}{\text{Volume of soil } (cm^3)}$$

Weight of oven dry soil of sample A = 180.5-112.0= 68.5 g

Weight of oven dry soil of sample B = 169.6-107.0= 62.6 g,

Bulk density, A = 68.5/ 50.24 = 1.36g/cm^3

Bulk density, B = 62.6/50.24 = 1.25 g/cm^3

Results

Bulk density of soil samples are 1.36 and 1.25g/cm^3 respectively.

Fig. Method of collection of sample for bulk density measurement (Courtesy: Rao *et al.*, 2010b)

2. Weighing bottle method

Procedure

1. Weigh an empty R.D. bottle of about 50 mL capacity (without stopper)
2. Fill it with given soil, flush to the brim.
3. Tap the R.D. bottle about 20 times on the table or palm gently by letting it fall from a height of about 2-3 cm. This is an attempt to give natural packing to soil.
4. Fill the R.D. bottle with soil up to the brim again and weigh it.

5. Remove the soil from the R.D. bottle and then fill it with water by means of a burette and note down the exact volume of water needed to fill the R.D. bottle.

Observations

1. Weight of empty R.D. bottle (W_1) = ... g
2. Weight of R.D. bottle + soil (W_2) = ... g
3. Volume of water needed to fill the R.D. bottle (V) = ... c.c.

Calculation

1. Weight of soil = ... ($W_2 - W_1$) g
2. Volume of soil = ... (V) c.c.
3. Bulk density $= \frac{\text{Weight of bulk soil (g)}}{\text{Volume of bulk soil (c.c.)}}$
4. Bulk density (Mg m^{-3}) $= \frac{W2 - W1}{V}$
5. Bulk density (Mg m^{-3}) = --------------------------

Result

Bulk density of a given soil sample is .... Mg m^{-3}

■■■■

34

MEASUREMENT OF INFILTRATION RATE

Introduction

Infiltration refers to the process of entry of water from the surface of soil or rate at which water enters into the soil or the velocity at which water can seep into soil. It is dependent on soil type, soil structure (amount of aggregation) and soil water content. Soil crusting, compaction, vegetative cover, and root and earthworm channels also influence the ability of water to move through (infilt-rate) soil layers. For example, slow infiltration can be caused by increased soil compaction or loss of surface soil structure (reduced aggregation) through tillage. Leaving crop residues on the soil surface or increasing the organic matter content in the soil surface may maintain aggregation and enhance infiltration. Infiltration is rapid into large continuous pores at the soil surface and decreases as the size of these pores is reduced. Infiltration rate is commonly measured by the depth (mm) of the water layer that soil can absorb in an hour. Infiltration rate is much higher at the beginning of rain or irrigation and gradually decreases with time.

It is an important soil physical property deciding the rate of water application and time required to apply desired amount of water or storing water in the soil profile for plant growth (designing efficient irrigation method) and predicting runoff and erosion. Infiltration rates are best determined when the soil is at or near field capacity, usually 12 to 48 hours after the soil has been thoroughly wetted. Infiltration rate can be determined in the field using the double ring infiltrometer method based on ponding of water into a metal cylinder installed on the field surface and observing the rate at which water level is lowered in the cylinder. The infiltration rate decreases rapid initially and tends to approach a constant value after irrigation or rain. The nearly constant rate that develops after some time has elapsed from the start of irrigation is called the basic infiltration rate. Accumulated infiltration or cumulative infiltration is the total accumulated infiltration depth or total quantity of water that enters the soil in a given time. Infiltration rate and accumulated infiltration are the two parameters commonly used in evaluating the infiltration characteristics of soil. As farmers can influence the soil structure by means of cultural practices, they can also change the infiltration rate of their soil.

Materials

Galvanized iron cylinders 60 cm and 30 cm in diameter with circular caps (double ring infiltromter) and hammer, hook gauge, water, stop watch, graph paper, jute cloth and a spade or digging hoe.

Description of Double ring infiltrometer

- The equipment consists of two 25 or 30 cm height cylinders (concentric rings) made of smooth tough steel sheet (2 mm-rolled steel) of different diameter.
- Mostly the outer and inner cylinder have 60 and 30 cm diameter, respectively.

Procedure

- The spot at which the infiltration rate is to be determined is carefully cleaned and levelled.
- Drive the cylinder vertically downwards into the soil to a depth of about 10-15 cm by hammering gently on the circular cap of the cylinder without disturbing the soil.
- Remove the cap and tap the soil into the space between the soil column and the cylinder. Then drive the higher outer ring to the same depth as inner cylinder.
- Instead of directly striking the cylinder edges, a better option is to put a wooden plank on top of the cylinder and hitting the wooden plank slowly.
- The region between inner ring and outer ring is known as buffer pond.
- Place the hook gauge in the inner ring.
- Pond water first to the outer ring followed by inner cylinder to a height to 7-12 cm or level of water maintained in field during irrigation.
- When water is being applied into the inner ring, a piece of plastic, jute cloth or polythene is placed inside the ring to prevent any disturbance at or crusting of the soil surface. This sheet is subsequently removed and the initial reading of the water level is recorded immediately.
- The water within the two rings is to prevent a lateral spread of water from the infiltrometer.
- Record the recession in water level against time at certain time intervals and plot the curve.
- Initially the intervals between two consecutive observations are kept short, which become larger with passage of time.

- Repeat the process of pouring water and recession until two consecutive readings of infiltration rates are obtained or continue the test until the drop in water level is the same over the same time interval.
- Infiltration rate can be obtained from cumulative infiltration versus time curve.
- The infiltration curve can be obtained by plotting the points in rectilinear graph paper.
- The basic intake rate (BIR), which is a soil characteristic of the site is obtained from the graph in which infiltration rate (ordinate) has been plotted against elapsed time (abscissa).
- Note the infiltration capacity also from the same rectilinear graph by preparing a curve showing cumulative intake rate against time.

Precautions

- Drive the cylinder vertically downwards into the soil without disturbing the soil.
- Level the specific site selected for the study using a digging hoe or spade
- Care should be taken to ensure that the rings are pushed vertically downwards.
- Water depth in both the cylinders should be kept as the same depth during the observations.
- At least two infiltration tests should be carried out at a site to make sure that the correct results are obtained.

Reasons for decreasing the infiltration rate with time:

- Clogging of pores (micro pores)
- Swelling of particles reduces cross sectional area available for down ward movement of water
- Entry of air in soil.
- Lowering hydraulic gradient

Observations

S.No.	Elapsed time (min.)		Depth (cm)	Average Infiltration Rate (cm/hr.)	Cumulative Infiltration (cm)
	Initial reading (cm)	Final reading (cm)			

Result

Infiltration rate of soil is .... cm/ha.

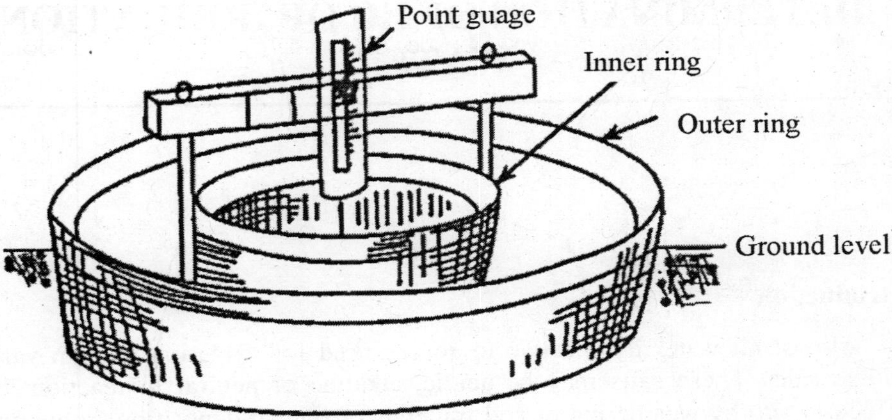

Fig. Double ring infiltrometer (Courtesy: Thomas *et al.*).

35

DETERMINATION OF EC OF IRRIGATION WATER

Introduction

Almost all water irrespective to source used for irrigation contain various soluble salts. These salts may be acidic, alkaline or neutral in reaction. Salts formed *in situ* by weathering of soil minerals or by salt deposition from applied water tend to accumulate in the soil profile. Salts like carbonates, bicarbonates, chlorides, sulphates and nitrates of Ca, Mg, Na, K, Fe and Mn are dissolved in natural water. Use of inferior water containing salts may cause deterioration of the physical properties of the soil and hinder crop growth. The high content of dissolved salts increases the osmotic pressure of water and influences osmo regulation of freshwater organisms. They reduce solubility of gases (like oxygen) and restrict utility of water for drinking, irrigational and industrial purposes. Quality of irrigation water is specially important in arid climates, where high rate of evaporation further contributes to the increased salt concentration in the soil and drainage water. The quality of the irrigation water is determined by the following chemical characteristics (Annexure 22) :

1. Total concentration of soluble salts, or salinity
2. Concentration of sodium relative to other cations, or sodicity
3. Anionic composition of water, particularly concentration of bicarbonates and carbonates
4. Concentration of boron, fluorine or other elements that may be toxic to plant growth

Electrical conductivity serves as a guide to know the extent of soluble salts (salinity) present in irrigation water. Electrical conductivity is the reciprocal of the electrical resistance or is the measure of the ability of the solution to carry an electric current. Quantitatively, electrical conductivity is the resistance in ohms, of a conductor, metallic or electrolytic, which is one cm long and has a cross-sectional area of the one cm^2. Hence, electrical conductivity is expressed as reciprocal of ohms/cm or mhos/cm. The terms, electrical conductivity and specific electrical conductance are synonymous. Electrical conductivity in terms of mhos/cm is such a large unit that most natural water has a value much less than one unit. For convenience, these values are reported as millimhos (EC × 10^{-3}) or micromhos (EC × 10^{-6}) at 25°C. The conductivity increases with increase in temperature. A practical means of providing temperature compensation is to introduce a resistance element into the bridge circuit that may change with temperature at the same rate as the solution under test. Electrical conductivity is now expressed as deci siemens per metre (dS/m) which is equal to one millimhos/cm. The electrical conductivity bridge consists of AC source Wheatstone bridge, or conductivity cell and a null indicating device. The EC measurement is affected by dissolved CO_2, turbidity, temperature and the nature of various ions and their relative concentration in water.

The effect of salts on crop growth is believed to be mainly osmotic. Hence it is related to the total concentration rather than the concentration of individual ionic species. Two methods are commonly used for determination of salt concentration:

1. Total dissolved salts (TDS) - This method is accurate in the case of water free of bicarbonates
2. Electrical conductivity (EC) – This method is commonly used. The electrical conductivity of the water sample is determined, and this is approximately proportional to the amount of salts present in the water. The quality characterization of irrigation water based on. electrical conductivity is given in Table 1.

Table 1. Quality characterization of irrigation water based on EC

Class	EC (dS m^{-1})	Quality characterization
1	1.5	Extra fresh, normal, excellent, very sweet, non-saline, very high quality
2	1.5-3.0	Fresh, slightly saline, good, sweet, low salinity, high quality
3	3.0-5.0	Brackish, moderately saline, fair, medium salinity, marginal quality
4	5.0-10.0	Saline, bad, low (poor) quality
5	10.0-15.0	Highly saline, very bad, very low quality
6	15.0	Excessively saline, very bad, extremely low (poor) quality

Reagents

0.01 N KCl solution : Dry a small quantity of AR grade KCl at 60°C for 2 hrs in a hot air oven, Weigh 0.7456 g of it and dissolve in freshly prepared distilled water and make it to 1L. This solution gives an electrical conductivity of 1.41 dSm^{-1} at 25°C.

Procedure

- Take the water sample in a clean beaker (100 ml).
- Rinse the conductivity cell with distilled water and then the sample under test.
- Switch on and warm it for 10 minutes and adjust the current by rotating the dial in such a way that the width of a shadow in cathode ray (magic eye) will be maximum.
- Push set knob (Temperature and cell constant corrections are adjusted on the conductivity meter if provided).
- Dip the electrode (conductivity cell) upto standard mark.
- Take N/100 KCl solution (0.7456 g KCl in 1 litre distilled water at 60) to calibrate the instrument that will be at 1.4118 dSm^{-1} at 25ºC. For this turn the range knob on 20.
- Remove the sample after pushing set knob.
- Wash the electrode at each test by distilled water and wipe by filter paper.
- Take reading of test sample by pushing cell read knob and multiply with cell constant specific for each conductivity cell to get specific conductance.
- Direct reading can be obtained in digital type of meters.

Results

Observed values of EC are multiplied by the cell constant (usually given on conductivity cell) and a temperature factor to express results at 25ºC. Value of conductivity usually varies by 2% per degree increase or decrease in temperature (Details are given in Annexure 23). Cell constant of a conductivity cell may also be calculated as follows:

$$\text{Cell constant} = \frac{1.4118}{\text{Observed conductivity of KCl solution}}$$

Interrelationship

Salt concentration, mg/l (ppm) = EC dSm^{-1} x 640*

(*increase with increase in EC)

Osmotic pressure (atm) = EC dSm^{-1} x 0.36

Table 2. Conversion factors for irrigation water quality (Annexure 20).

Component	To Convert	Multiply by	To Obtain
Water nutrient or TDS	mg/L	1.0	ppm
Water salinity hazard	1 dS/m	1.0	1 mmho/cm
Water salinity hazard	1 mmho/cm	1,000	1 μmho/cm
Water salinity hazard	ECw (dS/m) for EC <5 dS/m	640	TDS (mg/L)
Water salinity hazard	ECw (dS/m) for EC >5 dS/m	800	TDS (mg/L)
Water NO_3N, SO_4-S, B applied	ppm	0.23	lb per acre inch of water
Irrigation water	acre inch	27,150	gallons of water

■■■■

36

DETERMINATION OF pH IN IRRIGATION WATER

Introduction

The pH can be defined as the measure of acidity or alkalinity of a solution or negative logarithm of H^+ ion concentration. The pH scale is used to measure the degree of acidity or alkalinity. The nutrient availability is largely governed by soil pH or soil reaction. A pH meter is used to measure the exact value of pH. The pH meter is actually a voltmeter which measures the potential difference between two electrodes which share a common electrolyte. One electrode is the reference electrode with a constant reproducible potential and a measuring electrode which develops a potential depending upon H^+ ions concentration of solution or suspension.

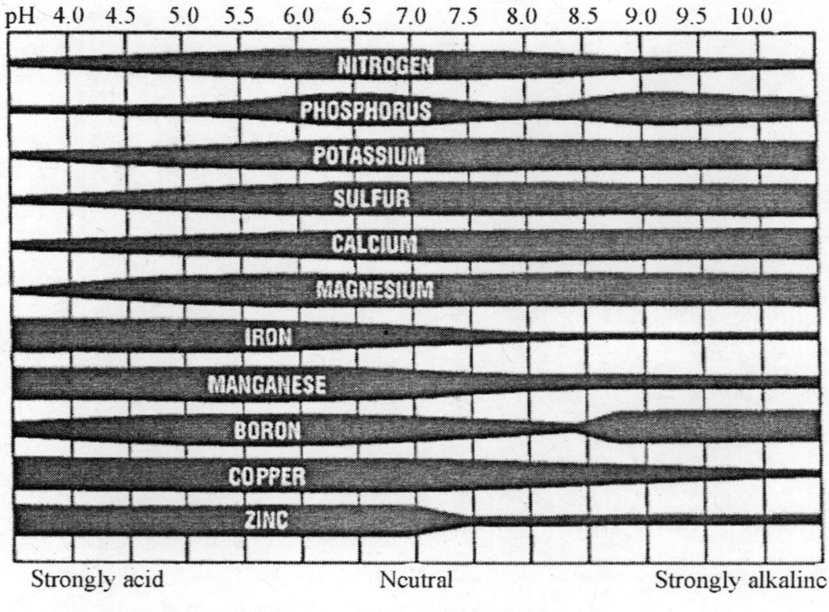

Fig. 1. The relationship of soil pH and nutrients availability.

$$pH = \log_{10} \frac{1}{(H^+)} = -\log_{10} (H^+)$$

pH scale

The pH scale extends from 0 to 14 with 7, the neutral point at which H^+ ions exist at the same concentration of OH^- ions. The minimum value of pH is 0, corresponding to the highest H^+ concentration and maximum value of 14 corresponding to the lowest H^+ concentration. Number greater than 7 indicates the degree of alkalinity and less than 7 indicate the degree of acidity. The concentration of H^+ ions at each successive value of pH is 10 times lower. H^+ concentration is measure in moles/litre.

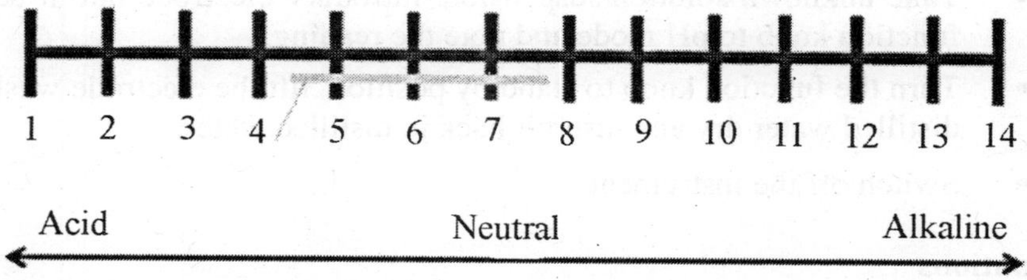

Fig. 2. Diagrammatic representation of pH scale.

Reagents

Standard buffer solution: Prepare buffer solution of pH 4.0, 7.0 and 9.2 using commercially available buffer tablets. Dissolve the respective tablets in freshly prepared distilled water and make up the volume to 100 mL. It is necessary to prepare fresh buffer solution after every few days, as these solutions should not be kept for long.

Procedure

A precise pH measurement of an unknown sample involves two measurements. The first measurement is of a buffer solution, the pH of which is known precisely at any given temperature, for the purpose of establishing a reference point. The second measurement is of the sample, with the pH value given relative to the buffer.

Steps

- Switch on the instrument to warm up for about half hour in stand by position.
- Adjust temperature at room temperature or temperature of solution by temperature compensation knob.

- Turn the function knob to pH mode to read the pH if necessary. Adjust the needles using the set buffer (standardise) knob to exact pH.
- Being back the selector switch to standby mode, lift electrode and wash it with distilled water, wipe it dry.
- Take 30 ml buffer pH 9.2 in the 50 ml beaker, introduce the electrode into it and turn the function knob to pH mode. The meter should read a pH of 9.2 (double buffer mode). If not the instrument/electrode/buffer is faulty
- Put selector switch to stand by position, lift electrode, wash and wipe by filter paper.
- Repeat standardise process two-three times to get exact reading.
- Take unknown solution/suspension. Introduce electrode into it turn the function knob to pH mode and note the reading.
- Turn the function knob to stand by position, lift the electrode wash with distilled water dry and insert it back in distilled water.
- Switch off the instrument

Precautions

- Before measuring the pH of unknown sample, instrument should be calibrated with standard solution of known pH (4.0, 7.0 and 9.2).
- Electrodes must be washed with jet of distilled water and dried gently with the help of filter paper or tissue paper before each measurement.
- The electrode of pH meter should always remain dipped or immersed in distilled water.
- The electrode of pH meter should not touch the beaker.
- Wait for 60 seconds for taking reading.

Result

pH of given sample of irrigation water is ....

■■■■

37

DETERMINATION OF CO_3^{2-} AND HCO_3^- IN IRRIGATION WATER

Introduction

Among the anions, CO_3^{2-} and HCO_3^- are found to affect crop growth adversely. Carbonates and bicarbonates in solution can be determined by titrating the solution with standard sulphuric acid (N/10) using phenolphthalein and later on methyl orange as indicators. When the colour of phenolphthalein disappears, it shows the conversion of carbonate into bicarbonate. Now methyl orange can be added which gives yellow colour. The change of colour from yellow to rose red is an indication of the neutralization of bicarbonates.

$$2Na_2CO_3 + H_2SO_4 = 2NaHCO_3 + Na_2SO_4$$

$$2NaHCO_3 + H_2SO_4 = Na_2SO_4 + 2H_2O + 2CO_2$$

Reagent

(i) Standard sulphuric acid (0.02 N): Dilute 2.8 ml of concentrate H_2SO_4 to 1 litre with distilled water. It will be approximately 0.1 N. Dilute 200 ml of this solution to make again 1 litre for getting 0.02N H_2SO_4. Standardise the solution by titration against standard alkali prepared separately.

(ii) Methyl orange indicator (Prepare 0.5% solution in 95% alcohol): Dissolve 0.5 g of methyl orange dye in 100 ml of 95% ethyl alchol.

(iii) Phenolphthalein indicator (Prepare 0.25% solution in 60% ethyl alcohol): dissolve 0.25 g of phenolphthalein powder in 100 ml of 60% ethyl alcohol.

Procedure

- Take 50 ml of water sample with pipette into 100 ml clean conical flask.

- Add 2-5 drops of phenolphthalein indicator. Pink colour shows the presence of carbonates.
- Titrate it against 0.02 N H_2SO_4, till the solution becomes colourless (at this stage carbonate is converted into bicarbonate). Record the reading.
- Add 2-3 drops of methyl orange indicator to the colourless into conical flask.
- Titrate with 0.02 N H_2SO_4 till the colour changes from yellow to rose red.
- Note the volume of 0.02 N H_2SO_4 consumed.
- Repeat the above two steps at least three times to get three concordant readings.

Observations

(1) Amount of 0.02 N acid required for 50 ml water in the presence of phenolphthalein = --ml

(2) Amount of acid required for the neutralization of bicarbonate solution in the presence of methyl red = -----ml

(3) Total acid required for the complete neutralization of carbonates (2 x) = --- ml

(4) Total acid required for the complete neutralization of bicarbonate originally present in water (y – x) = (because x is the amount of bicarbonate derived form carbonate)

Calculation

For CO_3^{-2}

$$\frac{\text{meq}}{\text{litre}} \text{ of } CO_3^{2-} = \frac{\text{Vol. of } H_2SO_4 \times \text{Normality of } H_2SO_4 \times 1000}{\text{Aliquot taken}}$$

$$\frac{\text{meq}}{\text{litre}} \text{ of } CO_3^{2-} = \frac{\dfrac{\text{Vol. of } H_2SO_4 \times \text{Normality of } H_2SO_4 \times 1000 \times}{\text{Equivalent weight}}}{\text{Aliquot taken}}$$

$$\frac{\text{mg}}{\text{litre}} \text{ of } CO_3^{2-} = \frac{\text{meq}}{\text{litre}} \text{ of } CO_3^{-} \times \text{Equivalent weight}$$

$$\frac{\text{mg}}{\text{litre}} \text{ or (ppm) of } CO_3^{2-} = \frac{\text{meq}}{\text{litre}} \text{ of } CO_3^{-} \times 30$$

(Eq. wt. of CO_3 is .30)

For HCO_3^-

$$\frac{\text{meq}}{\text{litre}} \text{ of } CO_3^{2-} = \frac{\text{Vol. of } H_2SO_4 \times \text{Normality of } H_2SO_4 \times 1000}{\text{Aliquot taken}}$$

$$\frac{\text{mg}}{\text{litre}} \text{ of } HCO_3^- = \frac{\text{meq}}{\text{litre}} \text{ of } HCO_3^- \times \text{Equivalent weight}$$

$$\frac{\text{mg}}{\text{litre}} \text{ or (ppm) of } HCO_3^- = \frac{\text{meq}}{\text{litre}} \text{ of } HCO_3^- \times 61$$

(Eq. wt. of HCO_3^- is 61)

Results

Amount of CO_3^{2-} = in given sample of irrigation water is ....

Amount of HCO_3^- in given sample of irrigation water is ....

38

DETERMINATION OF Ca AND Mg IN IRRIGATION WATER

Introduction

The hardness of water is understood to be a measure of the capacity of water for precipitating soap. Hardness is defined as a characteristic of water, which represents the total concentration of the alkaline earth metal cations of calcium and magnesium. Soap is precipitated chiefly by the calcium and magnesium ions commonly present in water, but may also be precipitated by ions of other polyvalent metals, such as aluminium, iron, manganese, strontium and zinc, and by hydrogen ions. expressed as calcium carbonate. The hardness may range from zero to hundreds of milligrams per litre in terms of calcium carbonate, depending on the source and treatment to which the water has been subjected. The Ca and Mg in irrigation water samples are estimated by Versenate (EDTA) method. This method is more accurate and determines both Ca and Mg in a single titration. Hence, this method is recommended for soil testing laboratories.

Principle

Ca, Mg and a number of other ions form stable complexes with Versenate (EDTA-disodium dihydrogen ethylene diamine tetra-acetic acid) at different pH values. This property is used for determination of Ca and Mg in the solution titrated with 0.01 N Versenate (EDTA) using Eriochrome black T dye as indicator at pH 10. The NH_4Cl and NH_4OH buffer is used to raise the pH. The end point is the change of colour from wine red to blue or bluish green. Ca and Mg in irrigation water can also be measured by AAS (Atomic Absorption Spectrophotometer).

Reagents

(i) Ammonium chloride – ammonium hydroxide (NH_4Cl-NH_4OH) buffer solution: Take 67.5 g of ammonium chloride in 570 ml of concentrated ammonium hydroxide and make up to one litre. Adjust the pH to 10.

(ii) Eriochrome black T (EBT) indicator: Take 100 ml of 95% ethanol and dissolve 4.5 g of hydroxyl amine hydrochloride in it. Add 0.5 g EBT to it and prepare the solution. Hydroxyl amine hydrochloride removes the interference of Mn by keeping it in lower valence state (Mn^{2+}). Prepare a fresh solution every month.

(iii) 0.01 N Versenate (EDTA) solution: Dissolve 2.0 g of EDTA in 100 ml distilled water to which 0.05g magnesium chloride is added and dilute to 1 litre. Standardize this solution with calcium chloride solution (0.01N).

(iv) 0.01 N $CaCl_2$ solution: Dissolve 0.5 g pure, AR grade $CaCO_3$ (oven dried) in 10 ml of 3.0 N HCl and make the volume to one litre.

Procedure

i. Pipette 10 ml of irrigation water sample in a 100 ml conical flask.

ii. Add 10-20 ml distilled water.

iii. Add 10 drops of NH_4Cl-NH_4OH buffer and 3-4 drops of Eriochrome black T indicator.

iv. Titrate this solution with 0.01 N EDTA with continuously stirring the solution till the colour changes from red to bright blue or green and no tinge of wine red colour remains behind.

v. Note down the volume of EDTA used.

Observations

S.No.	Volume of water taken	Burette readings			Volume of versenate solution used (ml)
		Initial	Final	Difference	

Calculations

$$\text{Concentration of Ca + Mg}\left(\frac{\text{meq}}{\text{litre}}\right)=\frac{\text{Vol. of EDTA used}\times\text{Normality of EDTA}\times 1000}{\text{Volume of sample taken}}$$

$$\text{Ca + Mg}\left(\frac{\text{meq}}{\text{litre}}\right)=\frac{(x)\times(0.01)\times 1000}{10}=10\,x$$

$$\text{Concentration of Ca + Mg}\left(\frac{\text{meq}}{\text{litre}}\right)=\frac{\text{Vol. of EDTA used}\times\text{Normality of EDTA}\times 32.96}{\text{Volume of sample taken}}$$

(Eq. Wt of Ca + Mg = 32.96)

Result: Ca + Mg in given sample of irrigation water is --------------- meq/litre.

Precautions

(i) Always keep a blank for comparison of colour change in actual titration.

(ii) Near to end point, EDTA should be added very slowly i.e. one drop in 10 seconds.

(iii) Any metal whose Yersenate complex is less dissociated than that of Ca and Mg will be preferentially complexes before Ca and Mg and thus will be included in titration. These ions include Sn, Cu, Zn, Mn, Fe. But irrigation water contains above elements in traces hence their interference, if any, can be overlooked.

■■■■

39

DETERMINATION OF CHLORIDE IN IRRIGATION WATER

Introduction

Chloride is essential to plants in very low amount. The quality of irrigation water is the most critical factor for development of salt affected soils. It can affect fertility status, performance of irrigation system and its longevity. Therefore, knowledge of irrigation water quality is critical to understand as to what management changes are necessary for long-term productivity. Chloride is a specific ion in irrigation water which can cause toxicity to sensitive crops at high concentration. The high chloride concentrations cause more problems when applied with sprinkler irrigation as sodium (Table 1). Chloride tolerance among different plants in order of increasing tolerance is bean, onion, carrot, lettuce, corn, potato, alfalfa, wheat, sorghum, sugar beet, barley.

Table 1. Chloride classification of irrigation water.

Chloride (ppm)	Effect on Crops
< 70	Generally safe for all plants.
70-140	Sensitive plants show injury.
141-350	Moderately tolerant plants show injury.
> 350	Can cause severe problems.

Principle

The determination of Cl^- is done by $AgNO_3$ (Mohr's titration) method, which is based upon the fact that in solution containing Cl and chromate (Cr). Silver reacts with all the Cl and precipitates before the reaction with chromate begins. The appearance of the brick-red colour of the silver chromate precipitate is the end-point of the titration.

Apparatus: Burette with stand, pipette, volumetric flask, beaker and weighing balance.

Reagents

1. Potassium chromate (K_2CrO_4) indicator (5%) solution: Dissolve 5 g of K_2CrO_4 in about 50 mL distilled water and add drop by drop 1N silver nitrate ($AgNO_3$) until a slight permanent red precipitate is formed. Filter and dilute to 100 mL volume with distilled water. With high purity analytical reagent, the indicator solution can be prepared directly.
2. Standard silver nitrate solution (0.01 N): Dry about 3.0 g silver nitrate in an oven at 105°C for 2 h, cool in a desiccator and store in a tightly stoppered bottle and dissolve 1.69 g of dried silver nitrate ($AgNO_3$) in distilled water and make the volume to one litre. Standardize it against standard NaCl solution and keep in amber coloured bottle away from light.
3. Standard NaCl solution (0.01 N): Dissolve 0.58 g dried sodium chloride in distilled water and make up to one liter with distilled water.

Procedure

1. Pipette out 10 mL of water sample in a 100 mL conical flask.
2. Add 4-5 drops of K_2CrO_4 indicator (making it dark yellow).
3. Titrate against the standard AgNO3 solution with continuous stirring till the first brick-red tinge appears.
4. Note the burette reading and calculate the chloride (meq/L).

Calculation

$$\text{Cl (meq/L)} = \frac{V1 \times N \times 1000}{V}$$

Where:

V1 = Volume of 0.01 N $AgNO_3$ titrated for the sample (mL)

N = Normality of $AgNO_3$ solution

V = Volume of water sample used for measurement (mL)

Result: Chloride in irrigation water is .....(meq/L)

In order to standardize the $AgNO_3$ solution used in the determination of Cl:

1. Titrate 10 mL 0.01N NaCl solution against 0.01 N $AgNO_3$ after adding 4 drops potassium chromate solution until a permanent reddish-brown color appears.
2. Take the reading and calculate $AgNO_3$ normality:

$$NAgNO_3 = \frac{10 \times NNaCl}{V\ AgNO_3}$$

Where:

$NAgNO_3$ = Normality of $AgNO_3$ solution

$VAgNO_3$ = Volume of $AgNO_3$ solution used (mL)

$NNaCl$ = Normality of NaCl solution.

■■■■

40

DETERMINATION OF SULPHUR IN IRRIGATION WATER

Introduction

Sulphur is an essential element for plant growth that is not commonly incorporated in fertilizers. It is measured in irrigation water to give an indication of possible deficiency problems. If the concentration is less than about 50 ppm, supplemental sulphate may need to be applied for good plant growth. Sulphur is Occurs in natural water and high concentration of Sulphate have laxative effect (enhances when sulphate consumed with magnesium). Problem of odour and corrosion in wastewater treatment due to its reduction to H_2S. A critical range of 10-13 mg/kg $CaCl_2$-extractable SO_4-S has commonly been reported for cereal (e.g., wheat, maize) and oilseed (e.g., mustard) crops (Tandon, 1991).

Principle

Sulphate ions are precipitated as barium sulphate ($BaSO_4$) in an acidic media (HCl) with Barium Chloride. The precipitate is dried, ignited and weighed and form it the amount of sulphate is calculated.

$$NaSO_4 + BaCl_2 = BaSO_4 + 2NaCl$$

Precipitation Method

Sulfate in water is determined normally by barium sulfate precipitation (Richards, 1954).

Apparatus

Mechanical shaker, Muffle furnace.

Reagents

i. Methyl Orange Indicator [4-$NaOSO_2C_6H_4N$: NC_6H_4/-4-N $(CH_3)_2$] 0.1%

Dissolve 0.1 g *methyl orange* indicator in 100 mL DI water.

ii. Hydrochloric Acid Solution (HC1), 1:1

Mix equal portions of *concentrated hydrochloric acid* with DI water.

iii. Barium Chloride Solution ($BaCl_2.2H_2O$), 1N

Dissolve 122 g *barium chloride* in DI water, and bring to 1-L volume with DI water.

Procedure

1. Put an aliquot of soil extract containing 0.05 to 0.5 meq SO_4-S into a 250 mL Pyrex breaker and dilute to 50 mL.
2. Add 1 mL 1:1 hydrochloric acid solution and 2-3 drops methyl orange, if the color does not turn pink, add some more 1:1 hydrochloric acid.
3. Put breakers on a hotplate, heat to boiling, then add 10 mL 1 *N* barium chloride solution in excess to precipitate SO_4 as barium sulfate.
4. Boil for 5 to 10 minutes, cover with a watchglass and leave no cool.
5. Filter solution through ashless filter paper, collect the barium sulfate precipitate on the filter paper, and then wash it several times with warm DI water until no trace of chloride remains. The presence of chloride in the filterate can be checked by $AgNO_3$ solution.
6. After washing, place filter paper with precipitate into a pre-weighed and dried procelain crucible and put in an oven at 105°C for 1 hour to dry.
7. Transfer crucible to a muffle furance heated to 550°C, and leave to dry ash for 2-3 hours.
8. Take crucible out of the muffle furnace, and place in a desiccator to cool, weigh crucible on an analytical balance, and take the reading, **t.**

Calculation

For precipitate of Sulfate in soil:

$$SO_4\text{-}S\,(meq/L) = \frac{t-b}{V} \times 8583.7$$

Where,

t = Weight of crucible + $BaSO_4$ precipitate (g)

b = Weight of empty crucible (g)

V = Volume of extract used for measurement (mL)

■■■■

REFERENCES

Agarwal, R.L. 2008. Seed Technology. Oxford & IBH Publishing Co. Delhi.

Ahlawat, I.P.S., Sharma, O.P. and Saini, G.S. 1998. Scientific Crop Production in India. Aman Publishing House, Meerut.

Anonymous 2010. Water Management Practical Manual. Department of Agronomy, Acharya N.G. Ranga Agricultural University Rajendranagar, Hyderabad – 500 030.

Anonymous 2011. Manual on Irrigation Water Management. Department of Agronomy CCS Haryana Agricultural University, Hisar-125004.

Anonymous. Practical Manual on Irrigation and Water Management. Department of Agronomy College of Horticulture, Vellanikkara, Kerala Agricultural University, Thrissur-680 656

Balasubramaniyan, P. and Palaniappan, S.P. 2016. Principles and Practices of Agronomy (2nd edition), Agrobios (India), Jodhpur.

Chhabra, A.K. 2006. Practical Manual of Floral Biology of Crop Plants. Deptt. of Plant Breeding, CCS HAU, Hisar.

Chopra, S.L., Kanwar, J.S. and Rakshit, A. 2013. Analytical Agricultural Chemistry, Kalyani Publishers Ludhiana.

College of Agriculture, CSK Himachal Pradesh Krishi Vishvavidyalaya, Palampur-176062 (India).

De, Gopal Chandra. 1989. Fundamentals of Agronomy. Oxford & IBH Publishing Co., New-Delhi.

Dhama, A.K. 1996. Organic Farming for Sustainble Agriculture. Agro Bios Publishers, Jodhpur.

Gupta, O.P. 2015. Weed Management: Principles and Practices (3rd edition), Agrobios (India), Jodhpur.

Gupta, O.P. 2016. Modern Weed Management (3rd edition), Agrobios (India), Jodhpur.

ICAR. 2010. Handbook of Agriculture (6th edition), Indian Council of Agricultural Research, New Delhi.

Jackson, M.L. 1973. Soil chemical analysis, Prentice Hall of India, Pvt. Ltd New Delhi.

Kanwar, J.S. 1976. Soil Fertility: Theory and Practice. (ed.) ICAR, New Delhi pp. 583.

Khare, D. and Bhale, M.S. 2011. Seed Technology. Scientific Publishers (India), Jodhpur.

Krishan Chandra, 2005. Organic Manures. Regional Centre of Organic Farming, Bangalore, pp. 46.

Kumar, S., Kumar, P., Dhindwal, A.S. and Sewhag Meena. 2011. Manual on Irrigation Water Management. Department of Agronomy, CCS Haryana Agricultural University, Hisar-125004

Majumdar, D.K. Irrigation water Management : Principles and Practices, PHI.

Michael, A.M. 1987. Irrigation - Theory and Practice. Vikas Publishing House Pvt. Ltd., New-Delhi.

Michael, A.M. and Ojha, T.P. 2012. Principles of Agricultural Engineering. Volume II. 4th Edition, Jain Brothers, New Delhi.

Mishra, R.D. and Ahmed, M. 1987. Manual on Irrigation Agronomy. Oxford & IBH Publishing Co. Pvt. Ltd., New-Delhi.

Panda, S.C. 2012. Modern Concepts and Advance Principles in Crop Production. Agrobios (India), Jodhpur.

Prasad, Rajendra. 2002. Text Book of Field Crops Production. ICAR, New Delhi.

Punia S.S, Samunder Singh, SS Pahuja and AS Dhindwal. 2010. Practical Manual – Principles and Practices of Weed Management, AGRON 305 and 403. Department of Agronomy, CCS Haryana Agricultural University, Hissar 75 pages.

Rana, S.S. and Kumar S. 2014. Practical Manual (Principles and Practices of Weed Management) Department of Agronomy, Forages and Grassland Management

Rani, P.L. and Srivivas A. Practical Manual ON WEED MANAGEMENT. ANGRAU Department of Agronomy. College of Agriculture, Rajendranagar, Acharya N.G Ranga Agricultural University, Hyderabad- 500 030.

Rao, V.S. 2000. Principals of Weed Science (2nd edition), Oxford and IBH Publishing Co., New Delhi.

Rao, V.P., Suneetha, K.B. and Hemalatha. 2010a. Irrigation water management. Department of Agronomy, College of Agriculture, Rajendranagar, Hyderabad, 148 pages.

Rao. V.P., Devi, K.B.S., Hemalatha, S., Prasad, P.V.N., Reddy, G.P., Ramana, A.V., Chandrasekhar, K., Reddy, M.S., Ammaji, P. & Mukandum, B. 2010b. Water Management Practical Manual (Agro. 201). Department of Agronomy, Acharya N.G. Ranga Agricultural University, Rajendranagar, Hyderabad – 500 030

Rathore, P.S. 2000. Techniques and Management of Field Crop Production, Agrobios (India), Jodhpur.

Reddy, S.R. 2012. Agronomy of Field Crops. Kalyani Books, New Delhi.

Reddy, S.R. 2012. Principles of Crop Production (4th edition). Kalyani Publishers, Ludhiana.

Reddy, T.Y. and Reddy, G.H.S. 2016. Principles of Agronomy (2nd edition). Kalyani Publishers, Ludhian.

Ryan, J., Estefan, G., and R. Abdul, 2012. Soil and Plant Analysuis Laboratory Manual. Scientific Publishers (India), Jodhpur.

Saraswat, V.N., Bhan, V.M. and Yaduraju, N.T. 2003. Weed Management, ICAR, New Delhi.

Singh, Chhidda; Singh, P. and Singh, R. 2003. Modern Techniques of Raising Field Crops. Oxford & IBH Publishing Co., New Delhi.

Singh, D., Chhonkar, P. K. and Dwivedi V.S. 2005. Manual on Soil Plant and water analysis. Westville Publishing House, New Delhi.

Singh, P. 2001. Essentials of Plant Breeding-Principles and Methods. Kalyani Publishing House, New Delhi.

Singh, S.S. and Singh, R. 2011. Principles and Practices of Agronomy, Kalyani Publishers, New Delhi.

Thomas, C.G., Prameela, P. and Menon, M.V. Practical Manual on irrigation and water Management (2+1) (Agro. 1203). Department of Agronomy, College of Horticulture, Vellanikkara, Kerala Agricultural University, Thrissur-680 656.

Yawalkar, K.S. and Agarwal J.P. 1977. Manures and Fertilizers. Agricultural Horticultural Publishing House. Nagpur.

Annexure - 1
ICAR AND OTHER INSTITUTE ADDRESS

ICAR / OTHER RESEARCH INSTITUTES

Agricultural and Processed Food Export Development Authority (APEDA), New Delhi- 110016

Bhabha Atomic Research Centre (BARC), Trombay, Mumbai

Central Agroforestry Research Institute (CAFRI), Jhansi, U.P.

Central Arid Zone Research Institute (CAZRI), Jodhpur-342003 (Rajasthan)

Central Avian Research Institute (CARI), Izzatnagar, Bareili-243122 (U.P.)

Central Coastal Agricultural Research Institute (CCARI), Ela, Old Goa-403402

Central Food Technological Research Institute (CFTRI), Mysore

Central Inland Fisheries Research Institute (CIFRI), Barrackpore- 700120 (W.B)

Central Institute for Brackish Water Aquaculture (CIBA), 75 Santhome High Road, RA Puram, Chennai-600028 (T.N.)

Central Institute for Cotton Research (CICR), P.B. No. 2, Nagpur-440010 (Maharashtra)

Central Institute for Fisheries Education, Mumbai-400061 (Maharashtra)

Central Institute for Fresh Water Aquaculture (CIFA), Kausalyaganga, Bhubneshwar-751002 (Orissa)

Central Institute for Research on Buffaloes (CIRB), Sirsa Road, Hisar- 125001(Haryana)

Central institute for Research on Cattle (CIRC), Grass farm Road, P.B. No. 17, Meerut Cant, Meerut- 250001 (U.P.)

Central Institute for Research on Goats CIRG), Makhdoom, Mathura-281122 (U.P)

Central Institute for Subtropical Horticulture (CISH), Rehmankheda, PO Kakori, Lucknow-227107 (U.P)

Central Institute for Women in Agriculture (CIWA), 1199 Baramunda, Bhubneshwar-751030 (Odisha)

Central Institute of Agricultural Engineering (CIAE), Nabi Bagh, Berasia road,Bhopal-462038 (MP)

Central Institute of Arid Horticulture (CIAH), Sri Ganganagar Highway, Bikaner-334006 (Rajasthan)

Central Institute of Cotton Technology (CICT), P.B. No. 16640, Mumbai – 400019 (Maharashtra)

Central Institute of Fisheries Technology (CIFT), Cochin-682029 (Kerala)

Central Institute of Post Harvest Engineering & Technology (CIPHET), Ludhiana-141004 (Punjab)

Central Institute of Temperate Horticulture (CITH), Old Air Field, Rangreth-190007 (J&K)

Central Island Agriculture Research Institute (CIARI), Andaman and Nicobar

Central Marine Fisheries Research Institute (CMFRI), PB. No. 1603, Tatpuram P.O. Kochi-682018 (Kerala)

Central Medicinal and Aromatic Plants (CIMAP), Lucknow

Central Plant Protection Training Institute (CPPTI), Hyderabad

Central Plantation Crop Research Institute (CPCRI), Kasargod-671124 (Kerela)

Central Potato Research Institute (CPRI), Shimla-171001 (H.P.)

Central Research Institute for Dryland Agriculture (CRIDA), Santhoshnagar,

Central Research Institute for Jute & Allied Fiber (CRIJAF), Nilganj, Barrackpore-700120 (W.B.)

Central Rice Research Institute (CRRI), Cuttak, 753006 (Orrisa)

Central Salt and Marine Chemical Research Institute (CSMCRI), Bhavnagar

Central Sheep and Wool Research Institute (CSWRI), Avikanagar- 304501 (Rajasthan)

Central Soil & Water Conservation Research & Training Institute (CSWCRTI), 218 Kaulagarh Road, Dehradun -248195 (Uttaranchal) (Now Indian Institute of Soil and Water Conservation IISWC)

Central Soil Salinity Research Institute (CSSRI), Kernal- 132001 (Haryana)

Central Tobacco Research Institute (CTRI), Rajamundry-533105(A.P.)

Central Tuber Crops Research Institute (CTCRI), Sreekariam, Thiruvanthapuram-695017 (Kerela)

Council of Scientific and Industrial Research (CSIR), New Delhi

Directorate of Cashew Research (DCR), Puttur-574202 Dakshina Kannada (KN)

Directorate of Coldwater Fisheries Research (DCFR), Saurabh Cottage, Bhimtal, Nainital-263136 (Uttaranchal)

Directorate of Floricultural Research (DFR), Pune, Maharashtra

Directorate of Groundnut Research (DGR), PB No. 5, Ivnagar Road, Junagarh-362001 (Gujarat)

Directorate of Information and Publication of Agriculture (DIPA), KAB-I, Pusa, New Delhi-110012

Directorate of Knowledge Management in Agriculture (DKMA), New Delhi

Directorate of Mushroom Research (DMR), Solan, H.P.

Directorate of Rapeseed and Mustard Research (DRMR), Bharatpur – 321303 (Rajasthan)

Directorate of Rice Research (DRR), Hyderabad-500030 (A.P.) (Now Indian Institute of Rice Research IIRR)

Directorate of Seed Research (DSR), P.B. No. 11, Village Kushmour, P.O. Kaithuli, Mau – 275901

Directorate of Soybean Research (DSR), Indore - 452017 (M.P.)

Directorate of Weed Science Research (DWR), Maharajpur, Adhartal, Jabalpur-482004 (M.P.)

FRI Forest Research Institute (FRI), Dehradun

ICAR Research Complex for NEH Region, Barapani-793103 (Meghalaya)

ICAR Research Complex of Eastern Region, Walmi Complex, PO Phulwari Sharif, Patna-801505 (Bihar)

ICM Institute of Cooperative Management (ICM), Ahmedabad

IGNOU Indira Gandhi National Open University (IGNOU), New Delhi

Indian Agricultural Research Institute (IARI), New Delhi-110012

Indian Agricultural Statistics Research Institute (IASRI), Library Avenue, Pusa, New Delhi -110012

Indian Council of Forestry Research and Education (ICFRI), Dehradun

Indian Grassland and Fodder Research Institute (IGFRI), Jhansi-284003 (U.P.)

Indian Institute of Agricultural Biotechnology (IIAB), Ranchi

Indian Institute of Farming Systems Research (IIFRS), Meerut-250110

Indian Institute of Horticultural Research (IIHR), Hassarghatta Lake post, Banglore-560089 (KN)

Indian Institute of Maize Research (IIMR), IARI, Pusa Campus, New Delhi -110012

Indian Institute of Millets Research (IIMR), Rajendranagar, Hyderabad-500030 (A.P.)

Indian Institute of Natural Resins & Gums (IINRG), Ranchi

Indian Institute of Oil Palm Research (IIOR), Near Jawahar Navodaya Vidyalaya Pedavegi-534450, West Godavari (A.P)

Indian Institute of Oilseed Research (IIOR), Rajendranagar, Hyderabad-500030 (A.P.)

Indian Institute of Pulses Research (IIPR), Kanpur-208024

Indian Institute of Soil Science (IISS), Nabi Bagh, Berasia Road, Bhopal-462038 (M.P.)

Indian Institute of Spices Research (IISR), PB.No. 1701, Marikunnu PO, Kozhikode-673012 (Kerela)

Indian Institute of Sugarcane Research (IISR), Lucknow-226602

Indian Institute of Vegetable Research (IIVR), PB No. 01, PO.Jakini, Sahansapuram, Varanasi- 221305 (U.P.)

Indian Institute of Water Management (IIWM), Chandra Sekharpur, PO SE Railway, Project Complex, Bhubneswar-751023 (Orissa)

Indian Institute of Wheat and Barley Research (IIWBR), PB No.158, Karnal -132001 (Haryana)

Indian Lac Research Institute (ILRI), Ranchi- 834010 (Jharkhand)

Indian Veterinary Research Institute (IVRI), Izzatnagar, Bareili-243122 (U.P)

INSA Indian National Science Academy (INSA), Bahadur Shah Zafar Marg, NewDelhi

Institute of Forest Genetics and Tree Breeding (IFGTB), Coimbatore

Jute Agricultural Research Institute (JARI), Barrackpore

Jute Technology Research Institute (JTRI), Kolkata

National Academy of Agricultural Research and Management (NAARM), Hyderabad-500030 (A.P.)

National Agricultural Cooperative Marketing Federation of India Ltd. (NAFED), New Delhi

National Agroforestry Research Institute, Near Pahujdam, Jhansi-284003 (UP)

National Bank of Agriculture and Rural Development (NABARD)

National Bureau of Animal Genetic Resources (NBAGR), P.B. No. 129, Kernal-132001(Haryana)

National Bureau of Fish Genetic Resources (NBFGR), P.O. Dilkusha, Lucknow-226002 (U.P)

National Bureau of Plant Genetic Resources (NBPGR), New Delhi-110012

National Bureau of Soil Survey and Land Use Planning (NBSSLUP), Nagpur-440010 (MH)

National Centre for Integrated Pest Management (NCIPM), LBS Building, Pusa campus, New Delhi -110012

National Council of Education Research & Training (NCERT), New Delhi

National Dairy Development Board (NDDB), Safadarjung Enclave, New Delhi

National Dairy Research Institute (NDRI), Karnal-132001(Haryana)

National Institute of Abiotic Stress Management (NIASM), Malegaon, Baramati – 413115 (Pune- Maharashtra)

National Institute of Agricultural Economics & Policy Research (NIAEPR), Library Avenue, Pusa, New Delhi-110012

National Institute of Agricultural Extension Management (MANAGE), Hyderabad

National Institute of Animal Genetics (NIAG), Karnal (Harayana)

National Institute of Biotic Stress Management (NIBSM), Raipur

National Institute of Research on Jute & Allied Fibre Technology (NIRJAFT), Ranikuthi, Kolkata -700120 (W.B.)

National Research Centre for Banana (NRCB), Thogamalai Main Road, Thayar PO Thiruchirapalli -620102 (TN)

National Research Centre for Citrus (NRCC), PB No. 464 Shankarnagar PO, Nagpur-440010 (Now Central Citrus Research Institute-CCRI)

National Research Centre for Equines (NRCE), Hisar – 125001

National Research Centre for Grapes (NRCG), P B No. 3, Manjri Farm Post, Sholapur Road, Pune-412307 (MH)

National Research Centre for Litchi (NRCL), Manchi House, Club Road, Muzaffarpur-842002 (Bihar)

National Research Centre for Mithun, Jharnapani, Medziphema (NRCM), Nagaland-500059 (A.P.)

National Research Centre for Mushroom (NRCM), Chambaghat, Solan-173213 (H.P)

National Research Centre for Onion & Garlic (NRCOG), Ragurunagar-410505, Pune (MH)

National Research Centre for Orchids (NRCO), Pakyang, Sikkim- 737106

National Research Centre for Seed Spices (NRCSS), Tabiji, Ajmer-305206 (Rajasthan)

National Research Centre on Camel (NRCC), Jobber, P.B. No. 07, Bikaner-334001 (Rajasthan)

National Research Centre on Meat (NRCM), Saidabad, CRIDA campus, Hyderabad-500059 (A.P.)

National Research centre on Pigs (NRCP), Panjabari Road, 6_{th} Mile, Guwahati-785037 (Assam)

National Research Centre on Plant Biotechnology (NRPB), Pusa campus, New Delhi-110012

National Research centre on Pomegranate (NRCP), Solapur, Maharashtra

National Research Centre on Yak (NRCY), Dirang, West Kameng - 790101 (Arunachal Pradesh)

National Seed Corporation (NSC), Bee Bhawan, Pusa, New Delhi

Sugarcane Breeding Institute (SBI), Coimbatore – 641007

Vivekanand Parvatiya Krishi Anusandhan Sansthan (VPKAS), Almora-263601 (Uttaranchal)

Water and Land Management Training Research Institute (WLMTRI), Hyderabad

Water Technology Centre (WTC), IARI, New Delhi.

Annexure - 2

INTERNATIONAL ORGANISATIONS

Asian Institute of Technology (AIT), P.O. Box 2734, Bankok, Thailand-10501

Asian Vegetable Research and Development Centre (AVRDC), P.O.Box 42, Shanhua, Taiwan 74199

Bangladesh Agricultural Research Council (BARC), Dhaka

Canadian International Development Agency (CIDA), Cuba

Centro International de Agricultural Tropical (International Centre for Tropical Agriculture) (CIAT), Coli, Columbia, South Africa

Centro International deMajormiento de Maizy Trigo (International Centre for Improvement of Maize, Wheat and Barley) (CIMMYT), Maxico.

Commonwealth Scientific and Industrial Research Organization (CSIRO), Near Perth, Australia

Consultative Group on International Agriculture Research (CGIAR), Washington (USA)

Food and Agriculture Organization (FAO), Rome, Italy

International Benchmark Soil Network for Agrotechnology Transfer (IBSNAT), Hawaii

International Centre for Agricultural Research for Dryland Areas (ICARDA), Syrea

International Centre for Research in Agroforestry (ICRAF), Nairobi, Kenya

International Crop Research Institute for the semi Arid Tropics (ICRISAT), Patancheru, Hyderabad

International Development Research Centre (IDRC), Canada

International Fertilizer Development Centre (IFDC), Muscale Shoala, Alobama (USA)

International Food Policy Research Institute (IFPRI), Washington DC (USA)

International Fund for Agriculture Development (IFAD), Rome

International Institute for Applied Systems Analysis (IIASA), Luxemberg, Vienna, Austria

International Institute for Tropical Agriculture (ITA), Ibadan, Nigeria

International Network on Soil Fertility and Fertilizer Evaluation of Rice (INSFFER), Manila, Philippines

International Plant and Genetic Resources (IBPGR), Italy

International Plant Genetic Research Institute (IPGRI),

International Potato Centre (Centro Internationally de la Papa) (CIP), Peru, Mexico

International Rice Research Institute (IRRI), Manila Philippines

International Service for National Agricultural Research (INSNAR), The Hague, Netherland

International Soil Reference and Information Centre (ISRIC), Netherland

International Water Management Institute (IWMI)

Land Use Research Institute (LURI), Aberdeen, Scotland

National Institute on Agricultural Science (NIAS), Japan

Potash and Phosphate Institute (PPIC), Canada

Tropical Pesticides Research Institute (TPRI), Arushe, Tanzania

West Africa Rice Development Association (WARDA), Manrovia, Liberia

World Meteorological Organization (WMO), Geneva, Switzerland.

Annexure - 3

STATE AGRICULTURAL UNIVERSITIES

Acharya NG Ranga Agricultural University, Rajendranagar, Hyderabad (A.P.)

Agriculture University Jodhpur, Mandor , Jodhpur-342004

Agriculture University Kota, Borkhere, Kota-324001

Aligarh Muslim University, Aligarh

Anand Agricultural University, Anand (Gujarat)

Assam Agricultural University, Jorhat (Assam)

Banaras Hindu University, Varanasi

Bidhan Chandra Krishi Vishwavidhalaya, Mohanpur, Nadia (WB)

Bihar Agriculture University, Sabour, Bhagalpur – 813210 (Bihar)

Birsa Agricultural University, Kanke, Ranchi (Jharkhand)

Central Agricultural University, P O Box No. 23, Imphal, (Manipur)

Centrral Institute of Fisheries Education, Mumbai – 400061

Ch. Sarwan kumar Himachal Pradesh Krishi Vishwavidyalaya, Palampur (HP)

Chandra Shekhar Azad University of Agriculture & Technology, Kanpur,(U.P.)

Chaudhary Charan Singh Haryana Agricultural University, Hisar (Haryana)

Dr YSR Horticultural University, Venkataramanagudem, West Godawari –53401 (AP)

Dr. Babasaheb Sawant Kokan Krishi Vidyapeeth, Dapoli, Ratnagiri, (Maharashtra)

Dr. Panjabrao Deshmukh Krishi Vidyapeeth, Krishinagar, Akola-444 104 (Maharashtra)

Dr. Yashwant Singh Parmar University of Horticulture & Forestry, Solan, Nauni-173 230 (HP)

Gobind Ballabh Pant University of Agriculture & Technology, Pantnagar- 263145, Udam Singhnagar (Uttaranchal)

Guru Angad Dev Veterinary And Animal Science University, Ludhiana –141004

Indian Agricultural Research Institute, New Delhi – 110012

Indian Veterinary Research Institute, Izatnagar, Berrreili - 243122 (UP)

Indira Gandhi Krishi Vishwavidyalaya, Krishak negar, Raipur-492012 (Chhattisgarh)

Jawaharlal Nehru Krishi Vishwa Vidyalaya, Jabalpur – 482004 (MP)

Junagarh Agricultural University, Junagarh-362001 (Gujarat)

Karnataka Veterinary, Animal and Fisheries Sciences University, Nandnagar, Bidar-585401 (Karnataka)

Kerala Agricultural University, Vellanikara, Trichur-680656 (Kerala)

Kerala University of Fisheries and Ocean Studies, Pananagad, Cochin – 682506

Kerala Veterinary and Animal Sciences University, Pookot, Wayanand – 673576

Lala Lajpat Rai University of Veterinary & Animal Sciences, Hisar (Haryana)

Maharana Pratap University of Agriculture & Technology, Udaipur-3133001(Rajasthan)

Maharashtra Animal & Fisheries Sciences University, Nagpur – 4400006

Maharastra Animal Science & Fishery University, Seminary Hillls, Nagpur- 440006 (Maharashtra)

Mahatma Phule Krishi Vidyapeeth, Rahuri-413 402 (Maharastra)

Manyawar Kansiram Ji University of Agriculture and Technology, Banda (UP)

Nagaland University (School of Agricultural Sciences and Rural Development), Modziphema - 797106

Nanaji Deshmukh Veterinary Science University, Jabalpur – 482001 (MP)

Narendra Deo University of Agriculture & Technology, Kumar Ganj, Faizabad-224 229 (UP)

National Dairy Research Institute, Kernal – 132001 (Haryana)

Navsari Agricultural University, Navsari (Gujarat)

Orissa University of Agriculture & Technology, Bhubneshwar-751003 (Orissa)

Prof. Jayashankar Talengana State Agricultural University, Rajendra nagar, Hyderabad

Punjab Agriculture University, Ludhiana -141004 (Punjab)

Rajasthan University of Veterinary and Animal Sciences, Bikaner – 334006 (Rajasthan)

Rajendra Agriculture University, Pusa, Samastipur-848 125 (Bihar)

Rajmata Vijayaraje Scindia Krishi Vishwavidyalaya, Gwalior – 474002 (MP)

Rani Laxmi Bai Central Agricultural University, Jhansi

Sam Higginbottom Institute of Agriculture Technology and Sciences, Naini, Allahabad – 211007

Sardar Ballabh Bhai Patel University of Agriculture & Technology, Modipuram, Meerut-250 110 (UP)

Sardar Krushinagar Dantiwada Agricultural University, Sardar Krushinagar, Banskantha-385506 (Gujarat)

Sher-E-Kashmir University of Agricultural Sciences & Technology, Jammu - 188812 (J&K)

Sher-E-Kashmir University of Agricultural Sciences & Technology, Kashmir -191121 (J&K)

Sri Venkateswara Veterinary University, Trupati – 517502 (AP)

Srikonda Telangana State Horticultual University, Rajendra nagar, Hyderabad

Swami Keshvanand Rajasthan Agriculture University, Bikaner-33400 (Rajasthan)

Tamilnadu Agricultural University, Coimbattor-641003 (Tamilnadu)

Tamilnadu Fishries University, Nagapattanam-611003 (TN)

Tamilnadu Veterinary & Animal Sciences University, Chennai-600051 (Tamilnadu)

U.P. Pandit Deen Dayal Upadhyaya Pashu Chiktsha Vigyan Vishwavidhayala evam Go Anushandhan Sansthan, Mathura, (U.P.)

University of Agricultural Sciences GKVK Bangalore-560065 (Karnataka)

University of Agricultural Sciences, Dharwad-580005 (Karnataka)

University of Agricultural Sciences, Raichur – 584102 (Karnataka)

University of Agricultural Sciences, Shimoha, Karnataka

University of Horticultural Sciences, Sector-60, Navanagar, Bagalkot – 587102

Uttar Banga Krishi Vishwavidyalaya, Pundibari, Cooch behar – 736 165 (WB)

UttarakhandUniversity of Horticulture and Forestry, Pauri, Uttarakhand

Vasantrao Naik Marathwada Agricultural University, Parbhani-431 402

Viswa Bharti (Palli Siksha Bhawan), P.O. Sriniketan – 731236

West Bengal University of Animal & Fishery Sciences, 68 K Sahani, Kolkata- 700037 (WB).

Annexure - 4

IMPORTANT ABBREVIATIONS/ACRONYMS RELATED TO AGRICUTURE

A3P-Accelerated Pulse Production Programme

ACIAR- Australian Council on International Agriculture Research

AGRIS -International Information System for Agricultural Science and Technology

AIBP- Accelerated Irrigation Benefit Programme

AICRP -All India Coordinated Research Project

AOAC- Association of Official Analytical Chemists

AP Cess Fund -Agricultural Product Cess Fund

APEDA- Agriculture & Processed Products Export Authority

APWSS -Asian Pacific Weed Science Society

ARIC -Agricultural Research Information Centre

ARS- Agricultural Research Services

ASA- American Society of Agronomy

ASRB -Agricultural Scientist Recruitment Board

ATIC- Agricultural Technology Information Centre

ATMA- Agriculture Technology Management Agency

BARC- Bhabha Atomic Research Centre

BHC -Benzene Hexa chloride

CAB- Commonwealth Agricultural Bureau

CCIS -Comprehensive Crop Insurance Scheme

CDB- Coconut Development Board

CIDA- Canadian International Development Agency

CSC- Central Seed Committee

CSSA- Crop Science Society of America

CVRC- Central Variety Release Committee

DAP- Di Ammonium Phosphate

DARE- Department of Agricultural Research and Education

DBT- Department of Bio Technology

DDP- Desert Development Programme
DDT- Dichloro Diphenyle Trichloro Ethane
DPAP -Drought Prone Area Programme
DPD- Directorate of Pulses Development
DST- Department of Science and Technology
EAS- Employment Assurance Scheme
EDB- Ethylene Di Bromide
FAI- Fertilizer Association of India
FAO- Food and Agriculture Organisation
FCI -Food Corporation of India
FDA- Food and Drug Adulteration Act
FSC- Farm Science Centre
FYM -Farm Yard Manure
GSFC -Gujarat State Fertilizer Corporation Ltd
HYVP- High Yielding Variety Programme
IAAP- Intensive Agriculture Area Programme
IADP- Intensive Agriculture District programme
IARC- International Agricultural Research Centre
IFFCO- Indian Farmers Fertilizers Corporation Ltd
IJO-International Jute Organisation
IRDP- Integrated Rural Development Programme
ISA Indian Society of Agronomy
ISOPOM- Integrated Scheme for Oilseeds, Pulses, Oilpalm and Maize
ISSS- Indian Society of Soil Science
ISSS- International Soil Science Society
ISTS- Indian Society of Tobacco Science
IWSS-Indian Weed Science Society
KRIBHCO- Krishak Bharti Corporation Limited
KVK- Krishi Vigyan Kendra
LCU- Lac Coated Urea
LEISA- Low External Input Sustainable Agriculture
LLP- Lab to Land Programme
MFAL- Marginal Farmers and Agricultural Labour Programme
MGREGS- Mahatma Gandhi Rural Employment Guarantee Scheme
NAAS- National Academy of Agriculture Science
NAFED- National Agriculture Cooperative Marketing Federation of India Limited
NAIP- National Agriculture Innovative Project
NARP- National Agricultural Research Project

NATP- National Agricultural Technology Project

NBB- National Bee Board

NCA -National Commission on Agriculture

NCIPM- National Centre for Integrated Pest Management

NCU- Neem coated Urea

NDDB- National Dairy Development Board

NET- National Eligible Test

NFL- National Fertilizer Limited

NFSM- National Food Security Mission

NGO- Non Government Organisation

NICRA- National Initiative on Climate Resilient Agriculture

NITI AYOG- National Institute Transforming India

NMOOP- National Mission on Oilseeds and Oilpalm

NPMCR- National Policy for Crop Residue Management

NREGA- National Rural Employment Guarantee Agency

NSKS- Neem Kernel Seed Extract

NWDPRA- National Watershed Development Project for Rainfed Area

PKVY- Paramparagat Krishi Vikas Yojna

PRA- Participatory Rural Appraisal

READY- Rural Entrepreneurship and Awareness Development Yojna

RKVY- Rastriya Krishi Vikas Yojna

RLEGP- Rural Landless Employment Guarantee Programme

RRA- Rapid Rural Appraisal

SAARC- South Asian Association for Regional Cooperation

SAMETI -State Agricultural Management and Extension Training Institute

SAU- State Agricultural Universities

SFDA- Small Farmers Development Agency

SHC- Soil Health Card Scheme

SIDA- Swedish International Development Agency

SRI -System of Rice Intensification

SSDC- State Seed Development Corporation

SSP- Single Super Phosphate

TRYSEM- Training of Rural Youth Self Employment Programme

UGC- University Grand Commission

UNDP- United Nations Development Programme

UNEP- United Nations Environment Programme

UNESCO- United Nations Educational Scientific and Cultural Organisation

UNICEF- United Nations International Children Emergency Fund

USAID- United States Agency for International Development

USDA- United State Department of Agriculture

WMO -World Meteorological Organisation.

Annexure - 5

CONVERSION FACTORS BETWEEN IMPORTANT PRIMARY AND SECONDARY AGRICULTURAL COMMODITIES

Commodity	Conversion factor
Rice (cleaned)	production 2/3 of paddy production
Cotton	
Cotton	Lint Production 1/3 of kapas production
Cotton	seed production 2/3 of kapas production 2 time of cotton lint production
Jute	
100 yards of Hessian	54 lbs of raw jute
4148 yards of Hesian	1 ton of raw jute
5.55 bales of raw jute (of 180 kg each)	
1 Ton of sacking 1.11 tons of raw jute	
6.17 bales of raw jute (of 180 kg each)	
1 Ton of Hesian 1.05 tons of raw jute	
Sacking etc 5.85 bales of raw jute (of 180 kg each)	
Groundnut	
Kernel to nuts in shell	70 per cent
Oil to nuts in shell	28 per cent
Oil to kernels crushed	40 per cent
Cake to kernels crushed	60 per cent
Sesamum	
Oil to seeds crushed	40 per cent
Cake to seeds crushed	60 per cent
Rapeseed and mustard	
Oil to seeds crushed	33 per cent
Cake to seeds crushed	67 per cent

Linseed

Oil to seeds crushed	33 per cent
Cake to seeds crushed	67 per cent

Castor seed

Oil to seeds crushed	37 per cent
Cake to seeds crushed	63 per cent

Cotton seed

Oil to seeds crushed	14 to 18 per cent
Cake to seeds crushed	82 to 86 per cent

Niger seed

Oil to seeds crushed	28 per cent
Cake to seeds crushed	72 per cent

Kardi (safflower) seed

Oil to seeds crushed	40 per cent
Cake to seeds crushed	60 per cent

Neem seed

Oil to kernels crushed	45 to 50 per cent
Cake to kernels crushed	50 to 55 per cent

Soyabean seed

Oil to soybean seed crushed	18 per cent
Meal to soyabean seed crushed	73 per cent
Hull from soybean seed crushed	8 per cent
Wastage from soyabean seed crushed	1 per cent

Sugar

Gur from Cane crushed 11.20 to 11.50 per cent

Crystal sugar from gur refined (Gur refineries) 62.5 per cent

Crystal suga from cane crushed (Cane factories) 10-20 per cent

Khandsari sugar (Sulphur & non sulphur) from standard gur refined 46 per cent

Molasses from cane crushed 4.0 to 4.5 per cent

Cane – Trash from cane harvested 8.0 to 12.0 per cent

Cashew nut

Cashew Kernel 25 per cent of cashew nuts

Annexure-6

CONVERSION OF NUTRIENTS AND THEIR FORMS

Organic matter= organic carbon x 1.724

(Organic matter contains 58 % organic carbon)

% protein content= % N x 6.24

NO_3%= N% x 4.54 and N%= NO_3% x 0.22

P_2O_5%= P% x 2.29 and P% = P_2O_5% x 0.44

K_2O%= K% x 1.20 and K%= K_2O% x 0.83

CaO%= Ca % x 0.40 and Ca%= CaO x 0.71

MgO%= Mg % x 1.63 and Mg% = MgO % x 0.61

Annexure-7
AGRICULTURE REVOLUTIONS IN INDIA

Type of revolution	Related to-
Green revolution	Food production (Wheat and Rice)
White revolution	Milk production
Blue revolution	Fish production
Grey revolution	Fertilizer production
Golden revolution	Fruit (apple production)
Red revolution	Tomato/ meet production
Pink revolution	Prawn production / onion production
Silver revolution	Egg/ poultry production
Yellow revolution	Oilseed (mustard production)
Round revolution	Potato production
Brown/ black revolution	Solar energy
Brown	Food processing

Annexure - 8

VARIOUS DAYS AND YEARS IN AGRICULTURE

Particular	Day on organized
Army medical Crop Establishment day	1st January
National Science Day	28th February
World Consumer Day	15th March
World forestry Day	21st March
World Day for Water	22nd March
World Meteorological Day	23rd March
World Health Day	7th April
World Heritage Day	18th April
World Earth Day	22nd April
International Biodiversity Day	22nd May
World No Tobacco Day	31st May
World Milk Day (FAO)	1st June
World Environment Day	5th June
World Day to Combat Desertification	17th June
World Agriculture Day	1st July
World Ozone Day	16th September
World Habitat Day	3rd October
World Food Security Day	6th October
World Food Day	16th October
Agriculture Education Day	3rd December
World Soil Day	5th December
Kisan Divas	23rd December

Annexure -9

ANTI-NUTRITIONAL SUBSTANCE PRESENT IN DIFFERENT CROPS LIMITS THEIR CONSUMPTION

Berseem	Coumarines
Castor	Ricin
Cotton	Gossypol
Fenugreek	Ad-libitum
Lathyrus	BOAA
Lucerne	Saponin
Napier	Oxalates
Oat/ Paragrass	Nitrates
Pearlmillet	Oxalic acid
Sorghum	Hydrocynic Acid/Prussic acid
Subabul	Mimosine
Soybean	Saponin / Trypsin inhibitors
Sesame	Oxalate/ phytes
Safflower	Bitter and cathartic

Annexure - 10

RHIZOBIUM STRAINS AND ASSOCIATED CROPS

Host Group	*Rhizobium* Species	Crops	N fix kg/ha
Pea group	*Rhizobium leguminosarum*	Green pea, Lentil, Lathyrus	62- 132
Soybean group	*R. japonicum*	Soybean	57- 105
Lupini group	*R. lupine orinthopus*	Lupinus	70- 90
Alfafa group	*R. melliloti*	Alfalfa, Fenugreek	100- 150
Beans group	*R. phaseoli*	Beans	80- 110
Clover group	*R. trifoli*	Clover, Berseem	130
Cowpea group	*Rhizobium* sp	Moong, Redgram, Cowpea, Groundnut, Glyricidia, Daincha, Sunhemp	57- 105
Cicer group	*Rhizobium* sp	Bengal gram	75- 117

Annexure - 11
IMPORTANT CONVERSIONS

Yield or rate

- 1 pound per acre = 1.121 kilograms per hectare
- 1 kilogram per hectare = 0.892 pounds per acre
- 1 ton per acre = 2.242 tons per hectare
- 1 ton per hectare = 0.446 tons per acre
- 1 bushel per acre = 1.121 quintals per hectare
- 1 quintal per hectare = 0.892 bushel per acre

Area

- 1 square mile = 2.59 square kilometers
- 1 square kilometer = 0.386 square miles
- 1 acre = 0.00405 square kilometers
- 1 square kilometer = 247.1 acres
- 1 acre = 0.405 hectares
- 1 hectare = 2.471 acres

Irrigation water

- 1 Cusec = 28.3 l
- 1 Cumec = 1000 l or 10^3
- 1 ha mm = 10,000 l or 10^4
- 1 ha cm = 100000 l or 10^5
- 1 ha m = 100,00,000 l or 10^7
- 1 acre inch = 3630 cubic feet
- 1 acre foot = 43560 cubic feet

Annexure - 12

RATING CHART FOR SOIL TEST VALUES OF PRIMARY AND SECONDARY NUTRIENTS

Nutrients	Rating		
	Low	Medium	High
Organic carbon (%)	< 0.50	0.50 – 0.75	> 0.75
Available N (Kg/ha)	< 280	280 – 560	>560
Available P (Kg/ha)	<10	10 -25	>25
Available K (Kg/ha)	< 120	120– 260	> 260
Available S (mg/Kg)	<10	10 -20	>20
Available Ca	<1.5	–	>1.5

Rating chart for soil test values of micro nutrients

Nutrients	Rating	
	Deficient	Sufficient
Available Boron (mg/Kg)	< 0.50	> 0.50
Available Zinc (mg/Kg)	< 0:60	> 0.60
Available Fe (mg/Kg)	< 4.50	> 4.50
Available Mn (mg/Kg)	< 2.0	> 2.0
Available copper (mg/Kg)	< 0.20	> 0.20
Available Molybdenum (mg/Kg)	< 0.20	> 0.20

Annexure - 13

ESSENTIAL PLANT NUTRIENTS AND RELATED INFORMATION

Nutrient	% in dry wt	Chemical symbol	Form taken by plants	Essentiality year by
Basic nutrients(3)				
Carbon	45	C	CO_2	1804, De Saussure
Hydrogen	6	H	H_2O	1804, De Saussure, 1882, Sachs
Oxygen	45	O	O_2, H_2O	1804, De Saussure, 1882, Sachs
Macro nutrients (6)				
Primary nutrients (3)				
Nitrogen	1.5	N	NO_3^-, NH_4^+	1804, De Saussure
Phosphorus	0.2	P	PO_4^{3-}, HPO_4^{2-}, $H_2PO_4^-$	1839, Liebig, 1861, Ville
Potassium	1	K	K^+	1866, Birner & Lucanus
Secondary nutrients				
Calcium	0.2	Ca	Ca^{2+}	1862, Stohmann
Magnesium	0.5	Mg	Mg^{2+}	1875, Boehm
Sulfur	0.1	S	SO_4^{2-}	1866, Birner & Lucanus
Micro nutrients (8)				
Iron	0.02	Fe	Fe^{2+}, Fe^{3+}	1843, Gris
Manganese	0.02	Mn	Mn^{2+}	1926, Sommer & Lipman

Zinc	0.003	Zn	Zn^{2+}	1922, McHargue
Copper	0.001	Cu	Cu_2^+	1925, McHargue
Boron	0.006	B	BO_3^{2-}, $B_4O_7^{2-}$	1926, Sommer & Lipman
Molybdenum	0.0002	Mo	MoO_4^{2-}	1939, Arnon & Stout
Chlorine	0.3	Cl	Cl^-	1954, Broyer, Carlton, Johnson, & Stout
Nickel	0.0003	Ni	Ni^{2+}	1987, Brown, Welch, & Cary

Annexure – 14

EVALUATION OF HERBICIDE APPLICATION: THE EUROPEAN WEED RESEARCH COUNCIL RATING SYSTEM

Weed response system:

Rating	% Weed control	Verbal description
1	100	Completely destroyed
2	99.0- 96.5	Very good control
3	96.5 – 93.0	Good control
4	93.0-87.5	Satisfactory
5	87.5 – 80.0	Justy satisfactory
6	80.0- 70.0	Unsatisfactory
7	70.0 -50.0	Poor
8	50.0- 1.0	Very poor
9	0.0	As untreated

Crop response system

Rating	% crop injury	Verbal description
1	0.0	No reduction or injury
2	1.0-7.0	Very slight discolouration
3	3.5-7.0	More severe but not lasting
4	7.0-12.5	Moderate and more lasting
5	12.5- 20.0	Medium and lasting
6	20.0- 30.0	Heavy
7	30.0-50.0	Very heavy
8	50.0-90.0	Nearly destroyed
9	100.0	Completely destroyed

Annexure - 15

BASIC CALCULATIONS FOR WATER MANAGEMENT

1. Calculation of area of regular shapes

(1) *Triangles*

Case 1. When the height and base are known

A = 0.5 x base x height = 0.5bh

Case 2: When the height is not known; but all the three sides.

A= $\sqrt{s(s-a)(s-b)(s-c)}$; where *s* is semi perimeter; a,b and c are sides

s (semiperimeter) = 0.5(a+b+c)

(2) *Squares and Rectangles*

A = *l w*; where *l*, length and *w*, width

(3) *Rhombuses and Parallelograms*

A = b h ; where *b*, base and *h*, height

(4) *Trapeziums*

A = 0.5 (b + a) h ; where b, base ; a, top and h, height

The top (a) is the side opposite and parallel to the base (b). In a trapezium only the base and the top run parallel.

(5) *Circles*

A $=\pi r^2$; where *r*, radius of the circle .

2. Surface areas of canal cross-sections

The most common shape of a canal cross-section is an "up-side-down" trapezium. The area is called the canal cross-section and has a trapezium shape.

Thus, the formula to calculate its surface is similar to the formula used to calculate the surface area of a trapezium.

Surface area of the canal cross-section = 0.5 (b + a) x h

where *b*, base = bottom width of the canal; top line, *a* = top width of the canal; *h*, canal depth = height of the canal (from the bottom of the canal to the top of the embankment)

3. Wetted perimeter

In a channel, the area in contact with water is the wetted perimeter. The area is also called the wetted canal cross-section. It also has a trapezium shape and the formula to calculate its surface area is:

Surface area of the wetted canal cross-section = 0.5 (base + top line) x water depth

$= 0.5\ (b + a_1) \times h_1$

where: base (b) = bottom width of the canal; top line (a1) = top width of the water level; water depth (h1) = the height or depth of the water in the canal (from the bottom of the canal to the water level).

Worked out example: Calculate the surface area of the cross-section and wetted cross-section. The following data are given: Canal cross-section : base (b) = 1.25 m; top line (a) = 3.75 m; canal depth (h)= 1.25 m. Wetted cross-section: base (b)= 1.25 m; top line (a_1)= 3.25 m; water depth (h_1)=1.00m.

Canal cross-section, A = 0.5 x (b + a) x h

$= 0.5 \times (1.25\ m + 3.75\ m) \times 1.25\ m = 3.125 m^2$

Wetted cross-section, $A = 0.5 \times (b + a_1) \times h$

$= 0.5 \times (1.25\ m + 3.25\ m) \times 1.00\ m = 2.25 m^2$

4. Surface area of a farm

It may be necessary to determine the surface area of a farmer's field for designing irrigation systems. For example, when calculating how much irrigation water should be given to a certain field, the size of the field must be known.

When the shape of the field is regular and has, for example, a rectangular shape, it should not be too difficult to calculate the surface area once the length of the field and the width of the field have been measured

Worked out example: (1)What is the area of the field in hectares, if the length of the field is 60 m and width of the field 40 m .

Area, A = length x width = 60 m x 40 m = 2400 m^2= 2400/10000= 0.24ha

(2) In the following case, the original field is ABCD. For the purpose of calculation of area, the field should be divided in to several regular areas (square, rectangle, triangle, etc.). In the following case, the area of the filed can be

calculated, if all the sides AB, BD_1, D_1D, DC, CC_1, C_1A, and C D_1 are known or can be measured.

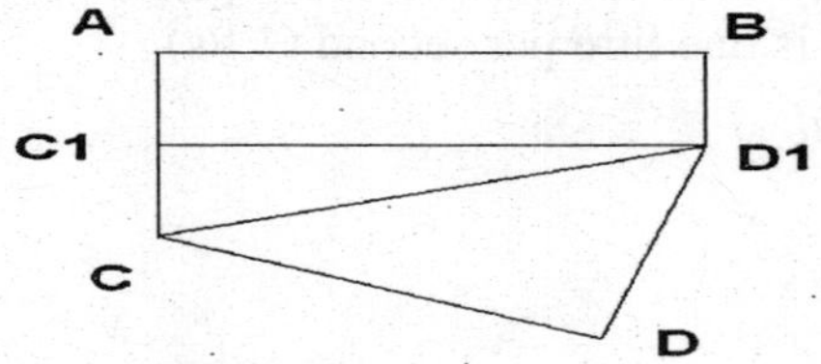

5. Volume of water

1 dm^3 = 1 litre and 1 m^3 = 1000 litres

Suppose there is a reservoir filled with water, with a length of 5 m, a width of 10 m and a depth of 2 m. All the water from the reservoir is spread over a field of 1 hectare. Calculate the water depth (which is the thickness of the water layer) on the field

As the first step, the volume of water must be calculated. It is the volume of the filled reservoir, calculated with formula

Volume (V) = length x width x height = 5 m x 10 m x 2 m = 100 m^3

As the second step, the thickness of the water layer is calculated using formula:

$$\text{water depth (d)} = \frac{\text{Volume of water (v)}}{\text{Surface area of the field (A)}}$$

Surface of the field = 10 000 m^2

Volume of water = 100 m^3

d= $100m^3/10000^2$= 0.01m = 10mm

Worked out example : A water layer of 1 mm thick is spread over a field of 1 ha. Calculate the volume of water (in m^3)

Volume of water (V) = Surface of the field (A) x Water depth (d)

Surface area of the field = 1ha= 10 000 m^2

Water depth = 1 mm =1/1 000 = 0.001 m

Volume (m^3)= surface of the field (m^2) x water depth (m)

V = 10 000 m^2 x 1/1 000 m = 10 m^3 = 10 000 litres

6. Flow-rate

The **flow-rate** of a river, or of a canal, is the volume of water discharged through the river, or the canal, during a given period of time. Related to irrigation, the volume of water is usually expressed in litres (l) or cubic metres (m^3) and the time in seconds (s) or hours (h). The flow-rate is also called **discharge-rate.**

For example, if the water running out of a tap fills a one litre bottle in one second, the flow rate (Q) is one litre per second (1 l/s).

Source: Thomas *et al.*

Annexure - 16

DISCHARGE THROUGH A 90° V NOTCH

Water head (cm)	Discharge (litres/ sec)	Water head (cm)	Discharge (litres/ sec)	Water head (cm)	Discharge (litres/ sec)
4.0	0.45	13	8.6	22.0	31.0
4.5	0.60	13.5	9.5	22.5	34.0
5.0	0.80	14	10.5	23.0	35.7
5.5	1.0	14.5	11.3	23.5	38.2
6.0	1.2	15	12.3	24.0	40.0
6.5	1.5	15.5	13.3	24.5	42.7
7.0	1.8	16	14.5	25.0	44.5
7.5	2.2	16.5	15.6	25.5	46.7
8.0	2.5	17	16.7	26.0	48.8
8.5	2.8	17.5	18.3	26.5	51.0
9.0	3.4	18	19.4	27.0	53.8
9.5	3.9	18.5	21.7	27.5	56.3
10.0	4.5	19	22.3	28.0	58.7
10.5	5.1	19.5	23.5	28.5	61.5
11.0	5.7	20	25.5	29.0	64.5
11.5	6.3	20.5	27.0	29.5	66.8
12.0	7.1	21	28.3	30.0	69.4
12.5	7.8	21.5	30.3		

Annexure - 17

FREE FLOW DISCHARGE VALUES FOR PARSHALL FLUMES

Water head (cm)	Discharge (litres/ sec) Throat width (cm)				Water head (cm)	Discharge (litres/ sec) Throat width (cm)			
	7.5cm	15 cm	23 cm	30cm		7.5cm	15 cm	23 cm	30cm
3	0.8	1.4	2.6	3.1	17	11.4	23.2	35.6	46.7
4	1.2	2.3	4.0	4.5	18	12.4	25.3	39.0	51.2
5	1.7	3.3	5.5	7.0	19	13.6	27.8	42.5	55.0
6	2.3	4.4	7.2	9.6	20	14.3	30.0	45.8	59.7
7	2.7	5.4	8.5	11.4	21	15.8	32.7	49.3	64.7
8	3.5	7.2	11.1	14.1	22	17.1	35.2	53.3	69.8
9	4.3	8.5	13.5	17.7	23	18.2	37.7	56.8	74.0
10	5.0	10.2	15.9	21.1	24	19.4	40.1	60.5	79.0
11	5.8	11.6	18.1	23.8	25	20.7	42.7	64.5	84.1
12	6.7	13.5	21.1	27.5	26	22.0	45.7	68.3	89.0
13	7.5	15.0	23.3	31.0	27	23.3	48.1	72.4	94.3
14	8.5	7.3	26.7	35.0	28	24.8	51.5	76.7	100.0
15	9.4	19.2	29.5	38.7	29	26.0	54.0	80.7	105.1
16	10.4	21.2	32.5	42.7	30	27.5	57.3	85.2	111.0

Annexure - 18

RELATIONSHIP BETWEEN THE VARIOUS UNITS OF SOIL WATER

The force by which the water is held by soil (soil water suction/tension) is the force that the plant roots have to overcome for extracting water retained by soil. It is expressed in the units of pressure. Earlier, it was generally expressed in bars (or atmospheres) but is now expressed in Pascal's (Pa). The relationship between the various units is given below:

1 bar = 1020 cm of water column* (Pressure exerted by a column of water 1020 cm high)

1 atm = 1030 cm of water column

1 bar = 10^5 Pa = 100 kPa = 0.1 MPa

Conventionally, the soil water potential is expressed in terms of hydraulic head i.e. soil water potential expressed in terms of height of water column. For example, if the soil water tension is 1 bar (soil water potential of -1 bar), in terms of water column height, it will be 1020 cm. For the sake of convenience, 1 bar is considered equivalent to 1 atm and approximately 1000 cm of water column height. Since these are fairly large numbers and handling such numbers is cumbersome, Schofield (1935) suggested the use of 'pF', which was defined as the logarithm of the negative pressure (tension or suction) head expressed in centimeters of water column. A tension head of 10 cm of water would be pF of 1, a tension head of 1000 cm would be pF of 3.

Annexure - 19

AVAILABLE WATER HOLDING CAPACITY OF SOILS OF DIFFERENT TEXTURAL CLASSES

Textured class	Available water (cm/cm of soil)
Sand	0.058
Loamy sand	0.074
Sandy Loam	0.146
Loam	0.191
Silt Loam	0.234
Sandy clay loam	0.209
Silty clay loam	0.204

Annexure - 20

CONVERSION OF % INTO PPM

%	=	ppm × 10^{-4}
1 %	=	10000 ppm
0.1%	=	1000 ppm
0.01%	=	100 ppm
0.001 %	=	10 ppm
0.0001%	=	1 ppm
1 mg/ litre	=	1 ppm
10 mg/ litre	=	10 ppm
1g/litre or 1000 mg/ litre	=	1000 ppm
1 lbs/Acre	=	ppm × 2

Annexure - 21

DEPLETION OF AVAILABLE WATER IN LIGHT AND HEAVY SOILS

Water retention curves for soil plotted in terms of per cent depletion of available water suggest that a coarse textured soil would get depleted of available water to a large extent even at a relatively lower water tension while a fine textured soil would retain a considerable portion of available water even at a relatively higher tension. Thus, the availability of water in a coarse textured soil would decrease further.

Suction (bar)	Per cent depletion	
	Light soil	Heavy soil
0.1	0	0
0.3	48	0
0.5	75	15
1.0	83	26
5.0	98	75

Annexure - 22

QUALITY PARAMETERS FOR IRRIGATION WATER

Class	EC(dS m^{-1})	SAR	RSC (meq/litre)
A - Good	<2.0	<10.0	<2.5
B - Normal	2-4	<10	<2.5
C - Sodic	<4	>10	>2.5
D - Marginally saline	4-8	>10	>2.5
E - Poor	>4	>10	>2.5

Annexure - 23

Temperature factors (f_t) for correcting resistence and conductivity data on soil extracts to the standard temperature of 25°C ($EC_{25}=EC_t \times f_t$). Where EC_{25}= EC at 25°C and EC_t= EC at room temperature

^0C	f_t	^0C	f_t	^0C	f_t	^0C	f_t	^0C	f_t
3.0	1.709	19.2	1.131	23.4	1.034	27.6	0.950	31.8	0.877
4.0	1.66	19.4	1.127	23.6	1.029	27.8	0.947	32.0	0.873
5.0	1.613	19.6	1.122	23.8	1.025	28.0	0.943	32.2	0.870
6.0	1.569	19.8	1.117	24.0	1.020	28.2	0.940	32.4	0.867
7.0	1.528	20.0	1.112	24.2	1.016	28.4	0.936	32.6	0.864
8.0	1.488	20.2	1.107	24.4	1.012	28.6	0.932	32.8	0.861
9.0	1.448	20.4	1.102	24.6	1.008	28.8	0.929	33.0	0.858
10.0	1.411	20.6	1.097	24.8	1.004	29.0	0.925	34.0	0.843
11.0	1.375	20.8	1.092	25.0	1.000	29.2	0.921	35.0	0.829
12.0	1.341	21.0	1.087	25.2	0.996	29.4	0.918	36.0	0.815
13.0	1.309	21.2	1.082	25.4	0.992	29.6	0.914	37.0	0.801
14.0	1.277	21.4	1.078	25.6	0.988	29.8	0.911	38.0	0.788
15.0	1.247	21.6	1.073	25.8	0.983	30.0	0.907	39.0	0.775
16.0	1.218	21.8	1.068	26.0	0.979	30.2	0.904	40.0	0.763
17.0	1.189	22.0	1.064	26.2	0.975	30.4	0.901	41.0	0.750
18.0	1.163	22.2	1.060	26.4	0.971	30.6	0.897	42.0	0.739
18.2	1.157	22.4	1.055	26.6	0.967	30.8	0.894	43.0	0.727
18.4	1.152	22.6	1.051	26.8	0.964	31.0	0.890	44.0	0.716
18.6	1.147	22.8	1.047	27.0	0.960	31.2	0.887	45.0	0.705
18.8	1.42	23.0	1.043	27.2	0.956	31.4	0.884	46.0	0.694
19.0	1.136	23.2	1.038	27.4	0.953	31.6	0.880	47.0	0.683